T0275632

The rapid progress in crystal growth and microfabrication technologies over the past two decades have led to the development of novel semiconductor devices. Among the most significant of these are resonant tunnelling diodes (RTDs), and this book is the first to give a comprehensive description of the physics and applications of these devices. The RTD, which utilises electron-wave resonance in double potential barriers, has emerged as one of the most important testing grounds for modern theories of transport physics, and is central to the development of new types of semiconductor nanostructure.

The opening chapters of the book set out the basic principles of coherent tunnelling theory and the various fundamental concepts necessary for the study of RTDs. Longitudinal-optical phonon-assisted resonant tunnelling, the effects of impurity scattering, femtosecond dynamics, non-equilibrium distribution, space charge build-up and intrinsic bistabilities are then described in detail. The applications of RTDs, such as in high-frequency signal generation, high-speed switching, and multi-valued data storage are reviewed, and the book closes with a chapter devoted to the new field of resonant tunnelling through laterally confined zero-dimensional structures.

Covering all the key theoretical and experimental aspects of this active area of research, the book will be of great value to graduate students of quantum transport physics and device engineering, as well as to researchers in both these fields.

Cambridge Studies in Semiconductor Physics
and Microelectronic Engineering: 2

Edited by

HAROON AHMED
Cavendish Laboratory, University of Cambridge
MICHAEL PEPPER
Cavendish Laboratory, University of Cambridge
ALEC BROERS
Department of Engineering, University of Cambridge

THE PHYSICS AND APPLICATIONS OF RESONANT TUNNELLING DIODES

TITLES IN THIS SERIES

1 Doping in III–V Semiconductors
E. F. Schubert

2 The Physics and Applications of Resonant Tunnelling Diodes
H. Mizuta and T. Tanoue

3 Electronic Transport in Mesoscopic Sysems
S. Datta

THE PHYSICS AND APPLICATIONS OF RESONANT TUNNELLING DIODES

HIROSHI MIZUTA AND TOMONORI TANOUE

Electron Devices Research Department
Central Research Laboratory
Hitachi Ltd

CAMBRIDGE UNIVERSITY PRESS
Cambridge, New York, Melbourne, Madrid, Cape Town, Singapore, São Paulo

Cambridge University Press
The Edinburgh Building, Cambridge CB2 2RU, UK

Published in the United States of America by Cambridge University Press, New York

www.cambridge.org
Information on this title: www.cambridge.org/9780521432184

First published 1995
This digitally printed first paperback version 2006

A catalogue record for this publication is available from the British Library

ISBN-13 978-0-521-43218-4 hardback
ISBN-10 0-521-43218-9 hardback

ISBN-13 978-0-521-03252-0 paperback
ISBN-10 0-521-03252-0 paperback

To Chihiro, Ryo and my parents

Hiroshi Mizuta

To Yoshiko, Marika, So and my parents

Tomonori Tanoue

Contents

Preface

The tremendous progress of crystal growth and microfabrication technologies over the last two decades has allowed us to explore a new field of semiconductor device research. The quantum mechanical wave-nature of electrons, expected to appear in nanometre-scale semiconductor structures, has been used to create novel semiconductor devices. The Resonant Tunnelling Diode (RTD), which utilises the electron-wave resonance occurring in double potential barriers, emerged as a pioneering device in this field in the middle of the 1970s. The idea of resonant tunnelling (RT) was first proposed by Tsu and Esaki in 1973, shortly after Molecular Beam Epitaxy (MBE) appeared in the research field of compound semiconductor crystal growth. Since then, RT has become of great interest and has been investigated both from the standpoint of quantum transport physics and also its application in functional quantum devices. Despite its simple structure, the RTD is indeed a good laboratory for electron-wave experiments, which can investigate various manifestations of quantum transport in semiconductor nanostructures. It has played a significant role in disclosing the fundamental physics of electron-waves in semiconductors, and enables us to proceed to study more complex and advanced quantum mechanical systems.

This book is designed to describe both the theoretical and experimental aspects of this active and growing area of interest in a systematic manner, and so is suitable for postgraduate students beginning their studies or research in the fields of quantum transport physics and device engineering. It facilitates the understanding of various aspects of the physics underlying observed complex characteristics by providing an appropriate theoretical basis at each stage. The book begins with a brief history of research on resonant tunnelling diodes (RTDs) in Chapter 1. Chapter 2 provides the basic global coherent tunnelling theory and various fundamental concepts needed for further study of RTDs. The major topics covered in Chapters 3 and 4 are LO-phonon assisted RT, RT via X-point states, the effects of impurity scattering, femtosecond dynamics, non-equilibrium distribution, space charge build-up and intrinsic bistabilities. Chapter 5 is intended as a review of various applications of RTDs such as high-frequency signal generation, high-speed switching, static memory, multi-valued memory, and signal processing. Chapter 6 is devoted to the highly current topic of RT through laterally confined zero-dimensional structures, including three-dimensional confinement effects, lateral mode non-conserving RT and RT via single impurity states.

We would like to express our special thanks to Professor H. Ahmed of the University of Cambridge, one of the editors of this new Cambridge series, who has given us the opportunity of writing the present book. We also gratefully acknowledge Dr K. Nakazato, Head of the Hitachi Cambridge Laboratory (HCL), Hitachi Europe Ltd, and Dr S. Takahashi of the Central Research Laboratory (CRL), Hitachi, Ltd (now of Hitachi Cable Ltd), for their continuous support and encouragement. The work described in the book would not have been possible without the help of many others. In particular, we would like to thank Dr C. J. Goodings of the Microelectronics Research Centre (MRC), University of Cambridge, for his invaluable contributions to the experimental aspects of the work described in this book. One of the authors, Hiroshi Mizuta, has collaborated with Dr Goodings for over four years on the project of zero-dimensional resonant tunnelling diodes and, in fact, some parts of Chapters 3, 4 and 6 have been taken from his PhD thesis entitled 'Variable-area Resonant Tunnelling Diodes Using Implanted Gates'. We are also indebted to Dr M. Wagner, now of HCL, and Professor C. Hamaguchi of Osaka University, for valuable discussions on the theoretical part of the work. We are grateful to Dr D. A. Williams, Dr R. J. Blaikie, Dr J. Allam and Dr J. White of HCL, and Dr J. R. A. Cleaver of MRC, for useful discussions. There are too many other friends and colleagues who have helped to complete this book to name individually. To all these, especially those at CRL and MRC, we would like to extend our heartfelt thanks.

Finally, we would like to express our gratitude to our wives and parents for their understanding, support and encouragement during the difficult times of the past few years while we completed this book.

Hiroshi Mizuta
Tomonori Tanoue

1
Introduction

1.1 Overview and background

Recent progress in crystal growth and microfabrication technologies have allowed us to explore a new field of semiconductor device research. The quantum-mechanical wave-nature of electrons is expected to appear in mesoscopic semiconductor structures with sizes below 100 nm. Instead of conventional devices, such as field effect transistors and bipolar transistors, a variety of novel device concepts have been proposed based on the quantum mechanical features of electrons. The *resonant tunnelling diode* (RTD), which utilises the electron-wave resonance in multi-barrier heterostructures, emerged as a pioneering device in this field in the mid-1970s. The idea of *resonant tunnelling* (RT) in finite semiconductor superlattices was first proposed by Tsu and Esaki in 1973 [1] shortly after molecular beam epitaxy (MBE) appeared in the research field of compound semiconductor crystal growth. A unique electron tunnelling phenomenon was predicted for an AlGaAs/GaAs/AlGaAs double-barrier heterostructure, based on electron-wave resonance, analogous to the Fabry–Perot interferometer in optics. In the particle picture, each electron is constrained inside the GaAs quantum well for a certain dwell time before escaping to the collector region. The bias dependence of the tunnelling current through the double-barrier structure shows negative differential conductance (NDC) as a result of RT. Experimental results [2] reported in the early days showed only weak features in current–voltage (I–V) characteristics at low temperatures and did no more than confirm the theoretical prediction of resonant tunnelling. This was because the MBE technique was still at an early stage and was not sufficiently developed to provide AlGaAs/GaAs/AlGaAs double-barrier structures in which the tunnelling current through a resonant state dominates over the background current through crystal defects. The rapid advances in MBE technique in the early 1980s, however, have led to a remarkable improvement in the NDC characteristics of the devices and have enabled us to exploit these to create new functional devices. An encouraging experimental result was reported in 1983 [3] about an observation of NDC at terahertz frequencies which has made this device very attractive for potential applications. The large NDC obtained, even at room temperature, leads directly to new multi-stable device operations, and novel three-terminal resonant tunnelling transistors [4]–[6] have been successfully developed which could take the place of conventional devices. In terms of both their high operating temperatures and voltages, resonant tunnelling devices are some of the most promising quantum mechanical devices available so far in the field of circuit research.

At the same time, the RTD has attracted people's interest purely from the standpoint of quantum transport physics. The RTD provides an enormous amount of information about the quantum mechanical aspects of electron transport in semiconductor microstructures. Resonant tunnelling can be described by the transmission and reflection processes of coherent electron waves through the double-barrier structure, as shown by Tsu and Esaki [1]. The tunnelling properties of electrons through the structure are simply expressed by using the transmission probability function, which can be obtained only by solving the Schrödinger equation with scattering boundary conditions. This simple picture of RT is called the 'global coherent tunnelling model' since the phase-coherence of an electron wave is maintained throughout the whole structure. In fact the global coherent tunnelling model describes what is going on in RTDs reasonably well, despite its simplicity. To be more realistic, however, inevitable scattering processes due to longitudinal-optical (LO) phonons and ionised impurities will more or less destroy the phase-coherence of electron waves. In these circumstances we have to go beyond the global coherent tunnelling model to deal with dissipative tunnelling processes. It has been found that scattering that breaks phase-coherence results in a broadening of the transmission peaks at resonance and a dramatic degradation of the peak-to-valley (P/V) current ratio of RTDs. In addition, frequent scattering processes may have an influence on the distribution of electrons in the quantum well, which changes from a completely ballistic one to a well-thermalised one depending on the ratio of typical scattering time to electron dwell time.

A proper theoretical description of the complicated transport properties beyond the global coherent tunnelling picture is given in terms of a non-equilibrium quantum transport theory such as the density matrix, Wigner function or Kadanoff–Baym and Keldysh techniques for the non-equilibrium Green's function. These non-equilibrium transport theories, however, have not been widely used for device modelling because of the difficulty in implementation. A numerically tractable description of quantum transport has recently been given using the density matrix and Wigner function [7]–[11]. The density matrix is a double-space–single-time function, obtained from the non-equilibrium Green's function by assuming translational invariance over time. The time evolution of the density matrix is then determined by the Liouville–von-Neumann equation, which is basically an equation of motion for the density matrix, with additional terms which represent collisional processes. Very recently, both the density matrix and Wigner function have been solved numerically in order to analyse electron transport in RTDs. The one-dimensional Liouville–von-Neumann equation has been solved in the finite-difference scheme, and some calculated results have been reported on the steady-state *I–V* characteristics and transient behaviour of RTDs. Although some major problems, such as boundary conditions and correct initial conditions for the density matrix and Wigner function, are still under discussion, the successful application of these approaches to electron devices is certainly a significant advance in quantum device simulation technology.

Several measurement techniques for RT have been introduced to investigate the complicated transport processes. Magnetotunnelling measurements have been widely used to investigate LO-phonon-mediated resonant tunnelling [12] and space charge build-up [13], [14]. Photoluminescence measurements have also been used for the determination of electron accumulation [15], [16]. One of the recent achievements of these techniques is that they have revealed that the RTD has an intrinsic bistability in the negative differential conductance region which stems from the dynamic redistribution of electrons in the quantum well. Tunnelling current measurement in the presence of hydrostatic pressure [17], [18] has been used in particular for the study of X-valley tunnelling. It has been shown that Γ–X-intervalley resonant tunnelling becomes significant for $Al_xGa_{1-x}As/GaAs/Al_xGa_{1-x}As$ structures with x greater than 0.45. A time-resolved photoluminescence technique with a picosecond laser has also been adopted [19] in order to study the dynamic aspect of resonant tunnelling. The RTD is a good laboratory for the investigation of quantum transport in semiconductor microstructures and, as well as being a negative conductance device, has played an important role in determining the physics of electron transport. In addition, a new idea for transport measurement, resonant tunnelling spectroscopy, [20] that utilises resonant tunnelling structure as an energy filter for electron waves has been proposed.

Since the late 1980s, the research on RTDs has proceeded in parallel with progress in lateral fabrication technology. Recent sophisticated nanometre fabrication technologies, such as electron beam lithography or focused ion beam implantation, have made it possible to fabricate an ultra-small RTD in which resonant tunnelling electrons are laterally confined in a small area of size below 100 nm [21], [22]. Such a device is often called a zero-dimensional (0D) RTD because electrons are confined laterally as well as vertically. The 0D RTD is a virtually isolated electron system which is weakly connected to its reservoirs, providing an ideal system with which to investigate electron-wave transport properties through 3D quantised energy levels. By using appropriate structural parameters such as barrier thickness, well width and size of lateral confinement, it is possible to realise a 'quantum dot' in which the number of electrons is well defined. Thus the 0D RTD has recently attracted a great deal of interest in single-charge-assisted transport (Coulomb blockade) [23]–[25]. Also, fine structure, attributable to a single ionised impurity in a dot, has recently been reported in this system [26]. The 0D RTDs now facilitate direct observation of phenomena originating from few-body problems which give rise to some of the most cumbersome aspects of quantum mechanical statistics.

This book provides a comprehensive and up-to-date study of resonant tunnelling in semiconductor heterostructures from the standpoints of both fundamental physics and applications to quantum functional devices. It is designed to describe this active and growing area of interest in a systematic manner and thus to be suitable for postgraduate students beginning their studies or research in the field of quantum transport physics and device engineering.

This introductory chapter provides a brief history of the last two decades of research on RTDs, as well as serving as a short overview of the topics in the following chapters.

Chapter 2 is intended as an introduction to the fundamental physics of RTDs. The idea of the global coherent tunnelling model is introduced to provide an intuitive and clear picture of resonant tunnelling. A theoretical formula for the transmission probability is presented based on the transfer matrix method, and this is then combined with a Tsu–Esaki formula to calculate the tunnelling current through double-barrier resonant tunnelling structures. In addition, a more analytical transfer Hamiltonian model is introduced since it is useful in the later discussion of more complicated tunnelling processes. The electron dwell time, which is one of the important quantities for the high-frequency performance of RTDs, is then studied. This is followed by discussions on RTDs with quantised emitter states and RT through multiple-well structures. Finally in this chapter, the idea of phase-coherence breaking scattering and associated incoherent resonant tunnelling is introduced. The problem of collision-induced broadening is then discussed in terms of the peak-to-valley (*P*/*V*) current ratio of the devices by using a phenomenological Breit–Wigner formula.

Chapter 3 deals with the physics of RT mediated by various elastic and inelastic scattering processes. Firstly, RT assisted by longitudinal-optical (LO) phonon emission, which leads to a shallow postresonant current peak is investigated. The theoretical analysis of this process is presented using the transfer Hamiltonian formula combined with the Fröhlich Hamiltonian. Magnetotunnelling measurements, used to extract detailed information about the interactions between tunnelling electrons and various types of phonon, are also discussed. Secondly, tunnelling processes through the upper X-valley, which become significant in $Al_xGa_{1-x}As/GaAs/Al_xGa_{1-x}As$ double-barrier structures with higher Al mole fractions, are discussed. Tunnelling current measurements under hydrostatic pressure, used to separate X-valley-related tunnelling from conventional Γ-valley tunnelling, are also presented. Finally, the effects of elastic scattering caused by residual background impurities or those diffused from heavily doped contact regions are studied by focusing on the resulting scattering-induced broadening of the transmission peaks.

Chapter 4 investigates the non-equilibrium distribution and femtosecond dynamics of electrons in RTDs. Dissipative quantum transport theory is introduced based on the Liouville–von-Neumann equation for the statistical density matrix. Numerical calculations are carried out in order to investigate dynamic space charge build-up in the quantum well which results in a unique intrinsic current bistability in the NDC region. Experimental studies of the charge build-up phenomenon are examined using magnetoconductance measurements and photoluminescence measurements. The effect of magnetic fields on intrinsic current bistability is also discussed.

Chapter 5 examines high-speed and functional applications of RTDs. Firstly, we study high-speed applications in the millimetre and submillimetre frequency ranges. In particular, high-frequency signal generation and high-speed switching are discussed. Secondly, functional applications of RTDs are described. The large negative differential conductance of RTDs allows new types of circuit to be designed based on different principles than those of conventional circuits. The functional applications are highly promising since RTDs can easily be integrated with conventional devices such as FETs and bipolar transistors. A one-transistor SRAM, multiple-valued memory cells using RTDs and signal processing circuits with significantly reduced numbers of devices are studied.

Chapter 6 is devoted to the study of resonant tunnelling through low-dimensional RTDs in which quantum size-effects are also expected in the lateral direction. Novel device structures fabricated using electron beam lithography and focused ion beam implantation are discussed, and the additional fine structure observed in the I–V characteristics of such devices is presented. Three-dimensional S-matrix theory is introduced to investigate multi-mode resonant tunnelling, and a numerical study of lateral mode non-conserving resonant tunnelling, which becomes important in the case of non-uniform confinement, is presented. A novel three-terminal device structure which enables the size of the quantum dot to be controlled is then used to study quantum size-effects fully. In addition, a newly found resonant tunnelling process, mediated by a single ionised impurity state, is discussed. Finally, the interplay of resonant tunnelling and the Coulomb blockade is discussed briefly.

1.2 References

[1] R. Tsu and L. Esaki, Tunneling in a finite superlattice, *Appl. Phys. Lett.*, **22**, 562, 1973.
[2] L. L. Chang, L. Esaki and R. Tsu, Resonant tunneling in semiconductor double barriers, *Appl. Phys. Lett.*, **24**, 593, 1974.
[3] T. C. L. G. Sollner, W. D. Goodhue, P. E. Tannenwald, C. D. Parker and D. D. Peck, Resonant tunneling through quantum wells at frequencies up to 2.5 THz, *Appl. Phys. Lett.*, **43**, 588, 1983.
[4] F. Capasso and R. A. Kiehl, Resonant tunneling transistor with quantum well base and high-energy injection: A new negative differential resistance device, *J. Appl. Phys.*, **58**, 1366, 1985.
[5] N. Yokoyama, K. Imamura, S. Muto, S. Hiyamizu and H. Nishi, A new functional resonant tunneling hot electron transistor (RHET), *Jpn. J. Appl. Phys.*, **24**, L853, 1985.
[6] A. R. Bonnefoi, T. C. Mcgill and R. D. Burnham, Resonant tunneling

transistors with controllable negative differential resistances, *IEEE Electron Device Lett.*, **EDL-6**, 636, 1985.
[7] W. R. Frensley, Simulation of resonant-tunneling heterostructure devices, *J. Vac. Sci. Technol.*, **B3**(4), 1261, 1985.
[8] J. R. Barker and S. Murray, A quasi-classical formulation of the Wigner function approach to quantum ballistic transport, *Phys. Lett.*, **93A**, 271, 1983.
[9] W. R. Frensley, Wigner-function model of a resonant-tunneling semiconductor device, *Phys. Rev.*, **B36**, 1570, 1987.
[10] N. C. Kluksdahl, A. M. Kriman, D. K. Ferry and C. Ringhofer, Self-consistent study of the resonant-tunneling diode, *Phys. Rev.*, **B39**, 7720, 1989.
[11] H. Mizuta and C. J. Goodings, Transient quantum transport simulation based on the statistical density matrix, *J. Phys.: Condens. Matter*, **3**, 3739, 1991.
[12] M. L. Leadbeater, E. S. Alves, L. Eaves, M. Henini, O. H. Hughes, A. Celeste, J. S. Portal, G. Hill and M. A. Pate, Magnetic field studies of elastic scattering and optic-phonon emission in resonant-tunneling devices, *Phys. Rev.*, **B39**, 3438, 1989.
[13] V. J. Goldman, D. C. Tsui and J. E. Cunningham, Resonant tunneling in magnetic fields: Evidence for space-charge build-up, *Phys. Rev.*, **B35**, 9387, 1987.
[14] M. L. Leadbeater, E. S. Alves, W. Sheard, L. Eaves, M. Henini, O. H. Hughes and G. A. Toombs, Observation of space-charge build-up and thermalization in an asymmetric double-barrier resonant tunnelling structure, *J. Phys.: Condens. Matter*, **1**, 10605, 1989.
[15] J. F. Young, B. M. Wood, G. C. Aers, R. L. S. Devine, H. C. Liu, D. Landheer, M. Buchanan, A. J. SpringThorpe and P. Mandeville, Determination of charge accumulation and its characteristic time in double-barrier resonant tunneling structures using steady-state photoluminescence, *Phys. Rev. Lett.*, **60**, 2085, 1988.
[16] M. S. Skolnick, D. G. Hayes, P. E. Simmonds, A. W. Smith, H. J. Hutchinson, C. R. Whitehouse, L. Eaves, M. Henini, O. H. Hughes, M. L. Leadbeater and D. P. Halliday, Electronic processes in double-barrier resonant tunneling structures studied by photoluminescence spectroscopy in zero and finite magnetic fields, *Phys. Rev.*, **B41**, 10754, 1990.
[17] E. E. Mendez, E. Calleja and W. I. Wang, Inelastic tunneling in AlAs–GaAs–AlAs heterostructures, *Appl. Phys. Lett.*, **53**, 977, 1988.
[18] D. G. Austing, P. C. Klipstein, A. W. Higgs, H. J. Hutchinson, G. W. Smith, J. S. Roberts and G. Hill, X- and Γ-related tunneling resonances in GaAs/AlAs double-barrier structures at high pressures, *Phys. Rev.*, **B47**, 1419, 1993.
[19] M. Tsuchiya, T. Matsusue and H. Sakaki, Tunneling escape rate of electrons from quantum wells in double-barrier heterostructures, *Phys. Rev. Lett.*, **59**, 2356, 1987.
[20] F. Capasso, S. Sen, A. Y. Cho and A. L. Hutchinson, Resonant tunneling spectroscopy of hot minority electrons injected in gallium arsenide quantum wells, *Appl. Phys. Lett.*, **50**, 930, 1987.
[21] M. A. Reed, J. N. Randall, R. J. Aggarwal, R. J. Matyi, T. M. Moore and A. E. Westel, Observation of discrete electronic states in a zero-dimensional semiconductor nanostructure, *Phys. Rev. Lett.*, **60**, 535, 1988.

[22] S. Tarucha, Y. Hirayama, T. Saku and T. Kimura, Resonant tunneling through one- and zero-dimensional states constricted by $Al_xGa_{1-x}As$/GaAs/$Al_xGa_{1-x}As$ heterojunctions and high-resistance regions induced by focused Ga ion-beam implantation, *Phys. Rev.*, **B41**, 5459, 1990.
[23] P. Guéret, N. Blanc, R. Germann and H. Rothuizen, Vertical transport in Schottky-gated, laterally confined double-barrier quantum well heterostructures, *Surface Science*, **263**, 212, 1992.
[24] M. Tewordt, L. Martin-Moreno, J. T. Nicholls, M. Pepper, M. J. Kelly, V. J. Law, D. A. Ritchie, J. E. F. Frost and G. A. C. Jones, Single-electron tunneling and Coulomb charging effects in asymmetric double-barrier resonant-tunneling diodes, *Phys. Rev.*, **B45**, 14407, 1992.
[25] Bo Su, V. J. Goldman and J. E. Cunningham, Single-electron tunneling in nanometer-scale double-barrier heterostructure devices, *Phys. Rev.*, **B46**, 7644, 1992.
[26] M. W. Dellow, P. H. Beton, C. J. G. M. Langerak, T. J. Foster, P. C. Main, L. Eaves, M. Henini, S. P. Beaumont and C. D. W. Wilkinson, Resonant tunneling through the bound states of a single donor atom in a quantum well, *Phys. Rev. Lett.*, **68**, 1754, 1992.

2

Introduction to resonant tunnelling in semiconductor heterostructures

This chapter is intended as an introduction to the fundamental physics of resonant tunnelling diodes (RTDs). The idea of *global coherent tunnelling* is introduced in order to provide an intuitive and clear picture of resonant tunnelling. The theoretical basis of the global coherent tunnelling model is presented in Section 2.2. The Tsu–Esaki formula, based on linear response theory, is adopted and combined with the *transfer matrix* method to calculate the tunnelling current through double-barrier resonant tunnelling structures (Section 2.2.1). The global coherent tunnelling model is improved by taking Hartree's self-consistent field (Section 2.2.2) into account. A more analytical transfer Hamiltonian formula is also presented (Section 2.2.3). Section 2.3 introduces the electron dwell time, which is one of the important quantities required to describe the high-frequency performance of RTDs. The effects of quantised electronic states in the emitter are then studied in Section 2.4. Section 2.5 describes resonant tunnelling through double-well structures. Finally, Section 2.6 discusses the idea of incoherent resonant tunnelling induced by phase-coherence breaking scattering. The problem of collision-induced broadening is then discussed in terms of the peak-to-valley (*P*/*V*) current ratio of RTDs by using a phenomenological Breit–Wigner formula.

2.1 Resonant tunnelling in double-barrier heterostructures

Let us start with a simple discussion of resonant tunnelling through the double-barrier heterostructure depicted in Fig. 2.1(a). A resonant tunnelling diode (RTD) typically consists of an undoped quantum well layer sandwiched between undoped barrier layers and heavily doped emitter and collector contact regions. The RTD is thus an open quantum

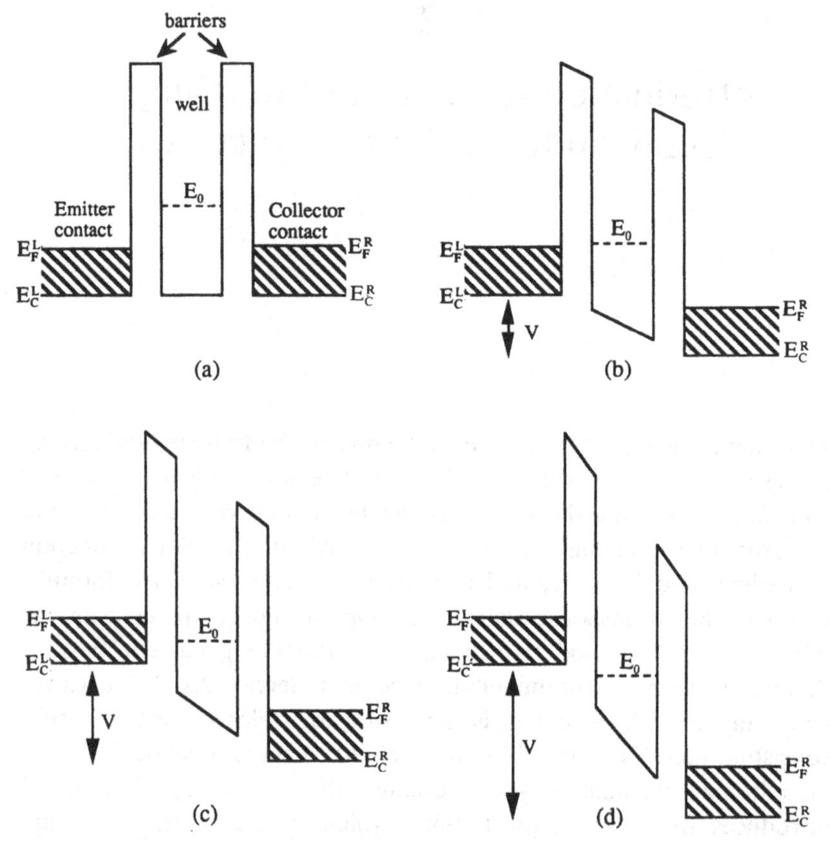

Figure 2.1 Conduction band profiles of a double barrier resonant tunnelling diode at four different bias states: (a) zero bias, (b) threshold bias, (c) resonance, and (d) off-resonance. Hatched regions represent the Fermi sea in the emitter and collector layers. E_F^L and E_F^R are the local Fermi energies in the emitter and collector, and E_C^L and E_C^R are the energies of the conduction-band edge in the emitter and collector.

system in which the electronic states are scattering states with a continuous distribution in energy space, rather than bound states with a discrete energy spectrum. Under these circumstances *quasi-bound states (resonant states)* are formed in the quantum well which accommodate electrons for a time that is characteristic for the double-barrier structure (the lowest state is denoted E_0 in the diagram).

The conduction-band diagrams for the RTD under four bias conditions are shown in Fig. 2.1(a)–(d): (a) at zero bias, (b) at threshold, (c) at resonance and (d) at off-resonance. The most common double-barrier system is based on the $Al_xGa_{1-x}As$(barrier)/GaAs(well)/$Al_xGa_{1-x}As$(barrier) heterostructure, and the relevant energy band is normally at the Γ-point, although the upper X-band takes part in the phenomenon to some extent for $Al_xGa_{1-x}As$ with $x > 0.45$ (Section 3.2). So-called 'resonant' tunnelling through the double-barrier structure occurs when the energy of the electrons flowing from the emitter coincides with the energy of the quasi-bound state, E_0, in the quantum well. The effect of the external bias, V, is to sweep the alignment of the emitter and quasi-bound states. Thus a resonant tunnelling current starts to flow when E_0 reaches the quasi-Fermi level, E_F^L, in the emitter (Fig. 2.1(b); this is hereafter called the threshold state of the RTD), reaches its maximum when E_0 passes through the Fermi sea in the emitter (Fig. 2.1(c): the resonant state) and ceases to flow when E_0 falls below the conduction-band edge in the emitter (Fig. 2.1(d): the off-resonant state).

In this simple form of tunnelling, both the energy and momentum parallel to the barriers are conserved since the double-barrier structure is entirely translationally invariant. In other words, the total energy of the electrons, $E(\mathbf{k})$, can be separated into lateral (x- and y-directions) and vertical (z-direction) components as follows:

$$E(\mathbf{k}) = \frac{\hbar^2(k_x^2 + k_y^2)}{2m^*} + E_z \tag{2.1}$$

where m^* is the electron effective mass. The lateral motion of the electrons is simply expressed in a plane-wave form with a lateral wavevector, $\mathbf{k}_{/\!/} = (k_x, k_y)$. In Chapter 3 we will show that this basic assumption is not necessarily correct when effects which break the lateral symmetry in the system are significant. In this chapter, however, this simple assumption will be adopted in order to aid intuitive understanding of resonant tunnelling.

The resonant tunnelling current through a double-barrier structure basically depends on the details of the transmission probability, introduced later in this section. Nevertheless, the expected shape of the I–V characteristics can be obtained from the following simple consideration. In the 3D emitter, at zero temperature, the electrons lie within a Fermi sphere of radius, k_F, which is the Fermi wave number in the emitter (see Fig. 2.2(a)). So long as tunnelling into the quantum well conserves the lateral wavenumbers, k_x and k_y, and the vertical energy, the electronic states involved in such a process can be represented by the interaction of the plane $k_z = q_R$ with the Fermi sphere, where q_R is the wavenumber

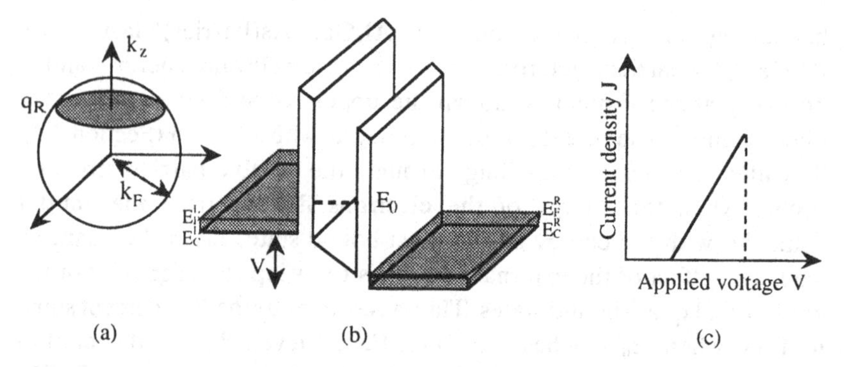

Figure 2.2 (a) A Fermi sphere of electrons in the emitter and (b) the corresponding schematic energy-band diagram. (c) Expected current–voltage (I–V) characteristics.

associated with the energy of the resonant state relative to the conduction-band edge:

$$q_R = \frac{\sqrt{(2m^*(E_0 - E_C^L))}}{\hbar} \tag{2.2}$$

where E_0 is the energy of the resonant state in the quantum well, and E_C^L is the energy of the conduction-band edge in the emitter (see Fig. 2.2(b)). The tunnelling current density will then be proportional to the density of the states indicated by the grey intersecting circle in Fig. 2.2(a). Therefore, as long as the transmission probability through the resonant state is nearly constant through the range of applied bias, the tunnelling current can be expressed as follows:

$$J \propto \pi(k_F^2 - q_R^2) \propto (E_F^L - E_0) \tag{2.3}$$

where E_F^L is the local Fermi energy in the emitter defined as $E_F^L = E_F + E_C^L$ (see Fig. 2.2(b)). As $E_F^L - E_0$ is proportional to the applied bias, J increases linearly until E_0 falls below the conduction-band edge in the emitter; the triangular shape of the I–V characteristics is then obtained as shown in Fig. 2.2(c).

The triangular shape mentioned in the above simple discussion is indeed observed for the I–V characteristics of fabricated double-barrier RTDs. Figure 2.3 shows the typical I–V characteristics observed for an RTD in which the double-barrier heterostructure consists of a GaAs quantum well layer sandwiched between two thin AlAs barrier layers [1]–[4].

Table 2.1. Structural parameters of the double-barrier RTD Material 1

Layer	Thickness (nm)
n–GaAs ($N_D = 1.0 \times 10^{18}\ \mathrm{cm}^{-3}$)	20.0
undoped – GaAs	5.0
undoped – AlAs	1.5
undoped – GaAs	4.5
undoped – AlAs	1.5
undoped – GaAs	5.0
n–GaAs ($N_D = 1.0 \times 10^{18}\ \mathrm{cm}^{-3}$)	20.0

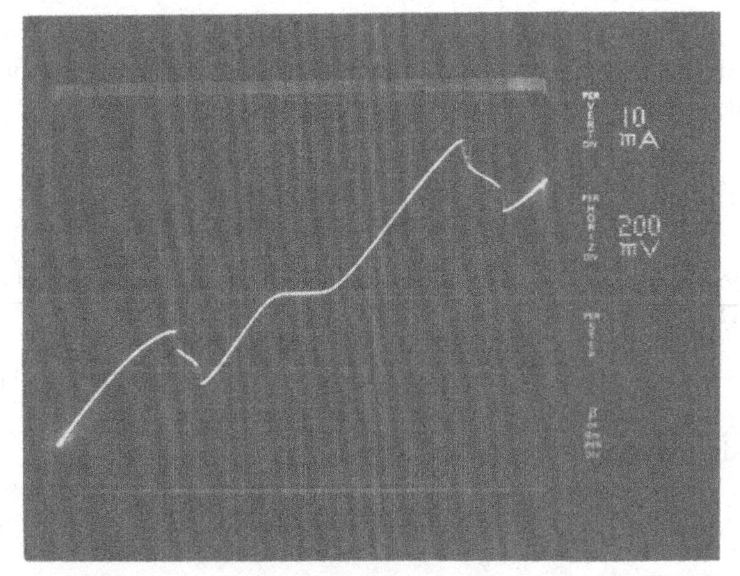

Figure 2.3 *I–V* characteristics for an AlAs/GaAs/AlAs double-barrier RTD with a mesa area of 16 μm^2 measured at 93 K in both forward- and reverse-bias directions.

The detailed geometry of the layer structure of the fabricated RTD is given in Table 2.1; this material is called *Material 1* in the following study. The *I–V* characteristic shown in Fig. 2.3 is found to reflect the basic nature of the resonance, although its details are quite different from those shown in Fig. 2.2(c).

An exact expression of the tunnelling current at finite temperatures is given by the following Tsu–Esaki formula:

$$J = J_{\rightarrow} - J_{\leftarrow} \tag{2.4a}$$

$$J_{\rightarrow} = 2 \sum_{k_x, k_y, k_z > 0} e v_z T(E_z) f_{\rm L}(\mathbf{k}) \{1 - f_{\rm R}(\mathbf{k})\}$$

$$= \frac{1}{2\pi^2} \int_0^{\infty} k_{/\!/} \mathrm{d}k_{/\!/} \int_0^{\infty} \mathrm{d}k_z e v_z T(E_z) f_{\rm L}(k_{/\!/}, k_z) \{1 - f_{\rm R}(k_{/\!/}, k_z)\} \tag{2.4b}$$

$$J_{\leftarrow} = 2 \sum_{k_x, k_y, k_z < 0} e v_z T(E_z) f_{\rm R}(\mathbf{k}) \{1 - f_{\rm L\leftarrow}(\mathbf{k})\}$$

$$= \frac{1}{2\pi^2} \int_0^{\infty} k_{/\!/} \mathrm{d}k_{/\!/} \int_{-\infty}^{0} \mathrm{d}k_z e v_z T(E_z) f_{\rm R}(k_{/\!/}, k_z) \{1 - f_{\rm L}(k_{/\!/}, k_z)\} \tag{2.4c}$$

where $T(E_z)$ is the *transmission probability function*, which will be explained in Section 2.2, and $f_{\rm L}(\mathbf{k})$ and $f_{\rm R}(\mathbf{k})$ are the Fermi distribution functions in the emitter and collector regions:

$$f_{\rm L,R}(\mathbf{k}) = \frac{1}{1 + \exp\left(\dfrac{E(\mathbf{k}) - E_{\rm F}^{\rm L,R}}{k_{\rm B}T}\right)} \tag{2.5}$$

where $E_{\rm F}^{\rm L} = E_{\rm F}^{\rm R} + V$ when the external voltage of V is applied to the RTD. By using

$$v_z = \frac{1}{\hbar} \frac{\mathrm{d}E_z}{\mathrm{d}k_z} \tag{2.6}$$

and integrating over k_x and k_y,

$$J = \int_0^{\infty} \mathrm{d}E_z T(E_z) S(E_z) \tag{2.7}$$

where $S(E_z)$ is the *electron supply function*, defined as follows:

$$S(E_z) = \frac{m^* e k_{\rm B} T}{2\pi^2 \hbar^3} \ln \left[\frac{1 + \exp\left\{\dfrac{1}{k_{\rm B}T}(E_{\rm F}^{\rm L} - E_z\right\}}{1 + \exp\left\{\dfrac{1}{k_{\rm B}T}(E_{\rm F}^{\rm R} - E_z\right\}} \right] \tag{2.8}$$

Near resonance the transmission probability function, $T(E_z)$, in general varies far more steeply than does $S(E_z)$. By simply replacing $T(E_z)$ by $\delta(E_z - E_0)$, it can easily be shown that eqns (2.7) and (2.8) reduce to expression (2.3) at zero temperature under large applied bias ($eV > E_{\rm F}$).

2.2 Theory of global coherent resonant tunnelling

This section introduces a simple theory of quantum transport which is based upon a *global coherent tunnelling* picture in which an electron does not experience any phase-coherence breaking events throughout the structure. First the *transfer matrix method* for 1D scattering problems is introduced in order to calculate the transmission probability function. The transfer matrix method, originally described by Tsu and Esaki, provides a means of calculating the transmission probability function, $T(E_z)$, numerically, and the maxima of this correspond to the quasi-bound states in the quantum well. The 1D Schrödinger equation is solved numerically with scattering boundary conditions on the wavefunctions, and the transfer matrices for the heterostructure are calculated. The Schrödinger equation is solved without including any scattering potentials except for the ideal multi-barrier potential resulting from the differences in electron affinity in heterojunction systems. The theory is first described in a non-self-consistent scheme in Section 2.2.1, neglecting the electron–electron interactions in RTDs. Section 2.2.2 then introduces the effects of space charge build-up in the quantum well on the global coherent tunnelling by adopting Hartree's self-consistent field model.

2.2.1 Transfer matrix theory of transmission probability

Let us start from the 1D, time-independent, effective mass Schrödinger equation:

$$H\Psi(z) = -\frac{\hbar^2}{2}\nabla\left(\frac{1}{m^*(z)}\nabla\right)\Psi(z) + V(z)\Psi(z) = E_z\Psi(z) \tag{2.9}$$

where $m^*(z)$ is the z-dependent conduction-band effective mass. The potential energy, $V(z)$, consists of the electron affinity and the self-consistent Hartree potential, which will be introduced in Section 2.2.2. In the transfer matrix method, the potential distribution of a resonant tunnelling structure is approximated by a series of small steps [5], as shown in Fig. 2.4. The wavefunction in the ith section can be expressed in plane-wave form:

$$\Psi^{(i)}_{k^{(i)}_z}(z) = A^{(i)}_{k^{(i)}_z}\exp(\mathrm{i}k^{(i)}_z z) + B^{(i)}_{k^{(i)}_z}\exp(-\mathrm{i}k^{(i)}_z z) \tag{2.10}$$

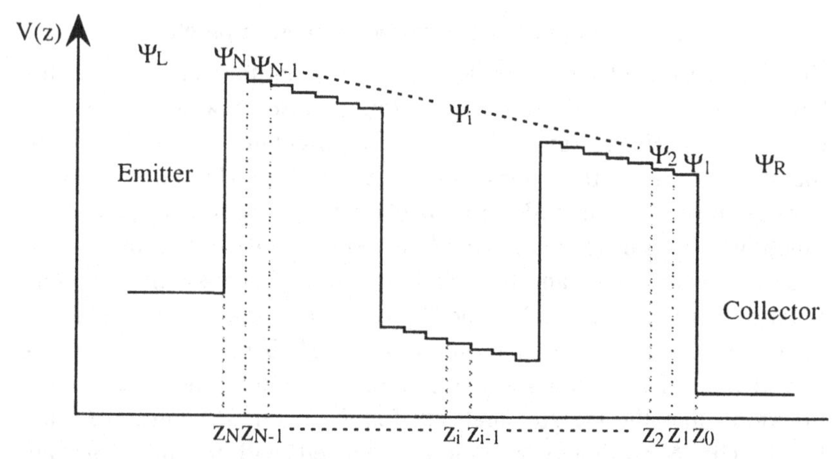

Figure 2.4 A series of small potential steps used for the transfer matrix calculations of a double-barrier resonant tunnelling structure.

where $k_z^{(i)}$ is the complex wavenumber defined as follows:

$$k_z^{(i)} = \frac{\sqrt{(2m^{*(i)}(E_z - V^{(i)}))}}{\hbar} \tag{2.11}$$

From the continuity of the probability flux of electrons, the following conditions on the wavefunctions hold at all boundaries:

$$\Psi^{(i)}_{k_z^{(i)}}(z_{i+1}) = \Psi^{(i+1)}_{k_z^{(i+1)}}(z_{i+1}) \tag{2.12a}$$

$$\frac{1}{m^{*(i)}} \left.\frac{\partial \Psi^{(i)}_{k_z^{(i)}}(z)}{\partial z}\right|_{z = z_{i+1}} = \frac{1}{m^{*(i+1)}} \left.\frac{\partial \Psi^{(i+1)}_{k_z^{(i+1)}}(z)}{\partial z}\right|_{z = z_{i+1}} \tag{2.12b}$$

The coefficients in the adjacent sections are then related to each other as follows:

$$\begin{pmatrix} A^{(i+1)}_{k_z^{(i+1)}} \\ B^{(i+1)}_{k_z^{(i+1)}} \end{pmatrix} = T^{(i)} \begin{pmatrix} A^{(i)}_{k_z^{(i)}} \\ B^{(i)}_{k_z^{(i)}} \end{pmatrix} \tag{2.13}$$

where the matrix $T^{(i)}$ is defined as

$$T^{(i)} = \begin{pmatrix} \alpha_+^{(i)} P & \alpha_-^{(i)}/Q \\ \alpha_-^{(i)} Q & \alpha_+^{(i)}/P \end{pmatrix} \tag{2.14a}$$

$$\alpha_{\pm}^{(i)} = \frac{1}{2}\{1 \pm (m^{*(i+1)}/m^{*(i)})\,(k_z^{(i)}/k_z^{(i+1)})\} \tag{2.14b}$$

$$P = \exp\{\mathrm{i}(k_z^{(i)} - k_z^{(i+1)})z_{i+1}\} \tag{2.14c}$$

$$Q = \exp\{\mathrm{i}(k_z^{(i)} - k_z^{(i+1)})z_{i+1}\} \tag{2.14d}$$

Thus the coefficients at the emitter and collector edges are connected by the transfer matrix T as follows:

$$\begin{pmatrix} A_{E_z}^{L} \\ B_{E_z}^{L} \end{pmatrix} = T\begin{pmatrix} A_{E_z}^{R} \\ B_{E_z}^{R} \end{pmatrix} \tag{2.15a}$$

$$T = T^{(N)}T^{(N-1)}T^{(N-2)}\ldots T^{(2)}T^{(1)} \tag{2.15b}$$

A complete set of wavefunctions is obtained by using the following scattering boundary conditions:

$$(A_{E_z}^{L}, B_{E_z}^{R}) = (1, 0) \tag{2.16}$$

for an incident electron wave from the emitter edge, and

$$(A_{E_z}^{L}, B_{E_z}^{R}) = (0, 1) \tag{2.17}$$

for that from the collector edge. The transmission probability $T(E_z)$ through the RTD is then given as follows:

$$T(E_z) = \frac{m^{*L}}{m^{*R}}\frac{k^{R}}{k^{L}}\frac{|A_{E_z}^{R}|^2}{|A_{E_z}^{L}|^2} \tag{2.18}$$

Numerical results for the energy dependence of the transmission probability at zero bias for Material 1 are shown in Fig. 2.5. The material

Table 2.2. Material parameters of $Al_xGa_{1-x}As$/GaAs heterostructures

1. *Electron effective mass m^**
$m^*(E) = 0.067(1 - 6\alpha(E - E_c)/E_g)m_0$; GaAs
$m^* = (0.067 + 0.083x)m_0$; $Al_xGa_{1-x}As$
Band gap $E_g = 1.42$ eV
Non-parabolicity parameter $a = -0.824$
2. *Conduction band discontinuity ΔE_c*
$\Delta E_c(x) = \beta \cdot 1.247x$ $(0 < x < 0.45)$
$\Delta E_c(x) = \beta \cdot (1.247x + 1.147(x - 0.45)^2)$ $(0.45 < x)$
Band parameter $\beta = 0.6$ (Miller's rule)
3. *Dielectric constant $\epsilon(x)$*
$\epsilon(x) = 13.1 - 3.0x$

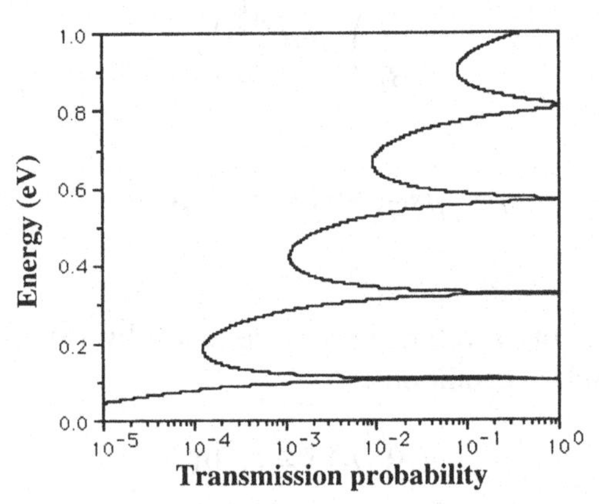

Figure 2.5 Transmission probability $T(E_z)$ versus energy E_z calculated for Material 1 under zero external bias.

constants of AlGaAs/GaAs heterosystems used in such calculations are summarised in Table 2.2. The effect of the non-parabolicity of the Γ-band on the effective mass in GaAs is taken into account [6] and the conduction-band offset value is assumed to be 60% of the Γ-point energy gap difference (Miller's rule). With increasing E_z, the transmission probability through the double-barrier structure tends to increase and reaches unity at several values of energy. This is because resonance occurs in the quantum well for certain wavelengths of incoming electrons, giving a unity transmission probability – the situation is analogous to the Fabry–Perot interferometer in optics.

Although the transmission probability can now be calculated numerically for any kind of multi-barrier structure, it is worth deriving

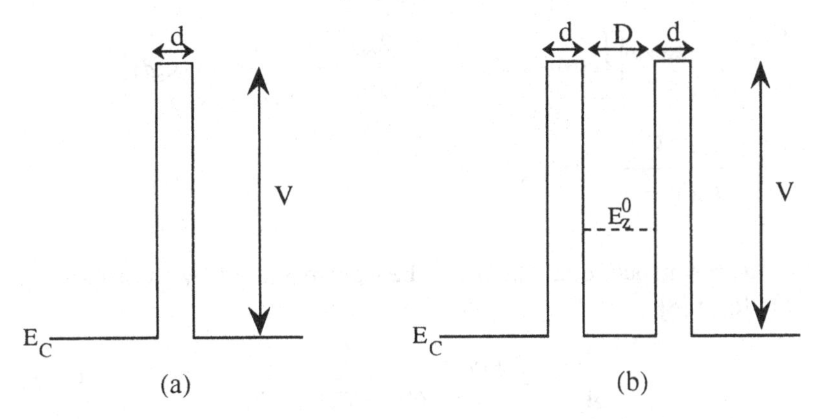

Figure 2.6 Simplified conduction band profiles of (a) single-barrier and (b) double-barrier structures at zero bias.

analytical expressions for some simple heterostructures in order to study the properties of $T(E_z)$ in further detail. First we consider the single-barrier structure with zero applied bias depicted in Fig. 2.6(a). By using eqns (2.13)–(2.18) it is easily shown that $A^3_{E_z}/A^1_{E_z}$ is given by the following expression:

$$\frac{A^3_{E_z}}{A^1_{E_z}} = e^{-ik_z d}\left\{\cosh(\kappa_z d) - i\frac{2E_z - V}{2\sqrt{(E_z(V-E_z))}}\sinh(\kappa_z d)\right\}^{-1} \tag{2.19}$$

and then the transmission probability through a single barrier, $T_{SB}(E_z)$, is as follows:

$$T_{SB}(E_z) = \left\{1 + \frac{V^2}{4E_z(V-E_z)}\sinh^2(\kappa_z d)\right\}^{-1} \tag{2.20}$$

where k_z and κ_z are defined as $\sqrt{(2m^*E_z/\hbar^2)}$ and $\sqrt{(2m^*(V-E_z)/\hbar^2)}$ respectively. If the energy E_z is much smaller than the energy barrier height, V, eqn (2.20) reduces to:

$$T_{SB}(E_z) \approx \frac{16E_z(V-E_z)}{V^2}e^{-2\kappa_z d} \tag{2.21}$$

Next, the transmission coefficient for a double-barrier structure [7] at zero bias (Fig. 2.6(b)) can be expressed in the following analytical form:

$$\frac{A^5_{E_z}}{A^1_{E_z}} = e^{-ik_z(2d+D)}\left[\left\{\cosh(\kappa_z d) - i\frac{2E_z - V}{2\sqrt{(E_z(V-E_z))}}\sinh(\kappa_z d)\right\}^2 e^{-ik_z D} + \frac{4V^2}{E_z(V-E_z)}\sinh^2(\kappa_z d)e^{ik_z D}\right]^{-1} \tag{2.22}$$

Close to resonance, eqn (2.22) can be approximated by the following simple form [8]:

$$\frac{A^5_{E_z}}{A^1_{E_z}} \approx e^{-ik_z(2d+D)}\frac{\Gamma}{(E_z - E^0_z) + i\Gamma} \tag{2.23}$$

where the energy, E^0_z, and width, Γ, of the resonant state are obtained approximately by solving the following equations:

$$2\cos(k^0_z D) + \frac{V - 2E^0_z}{\sqrt{(E^0_z(V - E^0_z))}}\sin(k^0_z D) \cong 0 \tag{2.24}$$

$$\Gamma \cong 4E^0_z e^{-2\kappa^0_z d}(k^0_z D + 2k^0_z/\kappa^0_z)^{-1} \approx 2\hbar T_{SB}(E^0_z)v^0_z/(D + 2/\kappa^0_z) \tag{2.25}$$

where v^0_z is the electron velocity ($=\hbar k^0_z/m^*$), and $T_{SB}(E^0_z)$ is the transmission probability through a single barrier (eqn (2.21)) at $E_z = E^0_z$ given by:

$$T_{SB}(E^0_z) \approx e^{-2\kappa^0_z d} \tag{2.26}$$

This simple expression ((2.23)) for the transmission coefficient is called the *Breit–Wigner formula* [9], which is a general form used to describe the resonant scattering spectrum and is widely used because of its simplicity (see also Sections 2.2.3 and 2.6).

2.2.2 Self-consistent calculations – introduction of space charge build-up

The theory described in the preceding section does not include any effects of electronic charge and the potential profiles have been approximated by the piecewise-linear model (see Fig. 2.1(a)–(d)). However, space charge distribution in RTDs, which reflects the probability density of electrons, gives rise to non-uniform potential profiles. Determination of the charge distribution in RTDs under such non-equilibrium conditions is one of the crucial issues under lively discussion, as it is closely related to the energy dissipation processes in the system that we study in Chapter 4. The energy dissipation processes of tunnelling electrons, such as LO-phonon emission, lead to a non-equilibrium electron distribution in the quantum well. Thus a simple Fermi–Dirac distribution function does not necessarily describe the accumulated electrons in the well and a complex non-equilibrium transport theory would in general be required; this issue will be studied in Chapter 4. In this section, we stay within the framework of the global coherent tunnelling model and thus express the electron density in RTDs as follows:

$$n(z) = 2 \sum_{k_z>0} |\Psi_{k_z}(z)|^2 f_{\mathrm{L}}(\mathbf{k}) + 2 \sum_{k_z<0} |\Psi_{k_z}(z)|^2 f_{\mathrm{R}}(\mathbf{k})$$

$$= \frac{k_{\mathrm{B}}T}{2\pi^2\hbar^2}\left(\int_0^{\infty} |\Psi_{k_z}(z)|^2 m^* \ln\left\{1 + \exp\left(\frac{E_{\mathrm{F}}^{\mathrm{L}} - E_z}{k_{\mathrm{B}}T}\right)\right\}\mathrm{d}k_z\right.$$

$$\left. + \int_{-\infty}^{0} |\Psi_{k_z}(z)|^2 m^* \ln\left\{1 + \exp\left(\frac{E_{\mathrm{F}}^{\mathrm{R}} - E_z}{k_{\mathrm{B}}T}\right)\right\}\mathrm{d}k_z\right) \quad (2.27)$$

The space charge distribution determined by eqn (2.27) results in *the self-consistent Hartree potential*, $V_{\mathrm{sc}}(z)$, which is obtained through the following Poisson equation:

$$\frac{\mathrm{d}}{\mathrm{d}z}\left(\epsilon(z)\frac{\mathrm{d}V_{\mathrm{sc}}(z)}{\mathrm{d}z}\right) = -e\{N_{\mathrm{D}}^{+}(z) - n(z)\} \quad (2.28)$$

The self-consistent potential is then added to $V(z)$ in the Schrödinger equation. The set of equations, (2.27) and (2.28), together with the Schrödinger equation, results in the well-known Hartree equation in which the exchange correlation between electrons is neglected. These equations can be solved numerically in an iterative way until the self-consistent solutions for $\Psi_{k_z}(z)$ and $V_{\mathrm{sc}}(z)$ are finally obtained [10]–[13].

Once we begin numerical calculations using the above self-consistent model, a question arises about the modelling of the heavily doped emitter and collector regions (and, if used, undoped spacer regions as well): should these regions be treated in the same way as the resonant tunnelling barrier region or not? As mentioned earlier, the present theory is based on an assumption that the electron waves propagate throughout the system without any phase-coherence breaking scattering. This is a plausible assumption as long as the theory is applied only for a thin, undoped, double-barrier structure, although it will be shown later that the effect of scattering processes on the coherent tunnelling picture often becomes significant. In the thick, heavily doped contact regions on both sides of the barrier structure, however, impurity scattering will occur frequently, along with some LO-phonon scattering. Consequently, these contact regions are frequently modelled using a classical Thomas–Fermi approximation in which electrons are assumed to occupy continuous energy states above the conduction-band edge and thus the electron density $n(z)$ in eqn (2.27) is given only by the difference between the quasi-Fermi level and the conduction-band edge. A fully quantum mechanical treatment of the contact regions can be made by adopting a dissipative quantum transport theory such as the density matrix theory or non-equilibrium Green's function theory, which we study in Chapter 4.

Examples of self-consistent solutions based on the above assumptions are illustrated in Fig. 2.7, which shows the equilibrium conduction-band profiles and electron density calculated self-consistently at room temperature for $Al_{0.26}Ga_{0.74}As/GaAs/Al_{0.26}Ga_{0.74}As$ double-barrier structures with well thicknesses of (a) 5 nm, (b) 6 nm and (c) 8 nm (the barrier thickness is fixed at 3 nm). Highly doped emitter and collector regions with a donor concentration of $1.0 \times 10^{18}\ \mathrm{cm}^{-3}$ are set adjacent to the barriers without intervening spacer layers. It can be seen that the conduction-band edge at the centre of the quantum well is bent upwards with increasing well width. This is due to the increasing electron population in the lowest quasi-bound state as its energy draws closer to the Fermi energy. The larger electron accumulation in the quantum well gives rise to a stronger self-consistent field resulting in larger band bending.

Self-consistent modelling has been applied to Material 1 (see Section 2.1): the *I–V* characteristics calculated at 77 K (solid black line) and 300 K (grey line) are shown in Fig. 2.8.

In Fig. 2.8 the *I–V* curve calculated using the non-self-consistent (piecewise-linear) model for the band diagram is shown in the inset in order to demonstrate the effect of self-consistent calculations of the band diagram. It can be seen that the resonant peak voltage obtained

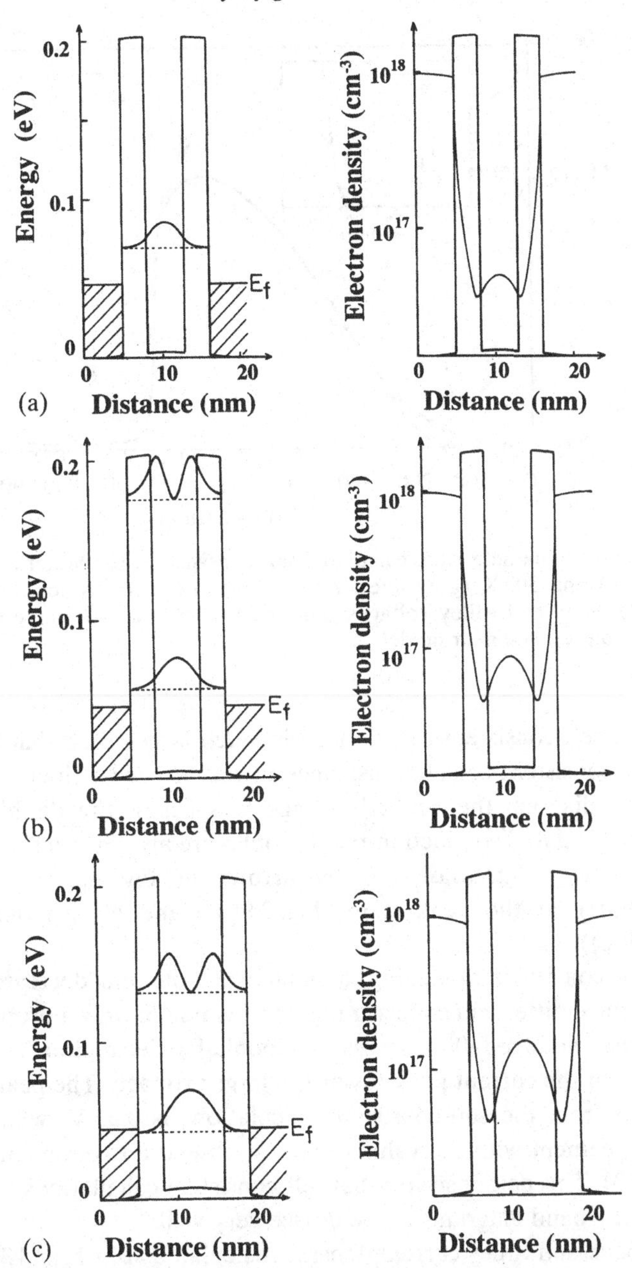

Figure 2.7 Self-consistently calculated energy-band diagrams and electron density of double-barrier RTDs with various well widths L_w: (a) 5 nm, (b) 6 nm, and (c) 8 nm. In the diagrams on the left the resonant energy levels and the probability density of electrons are also shown by dotted and solid curves respectively.

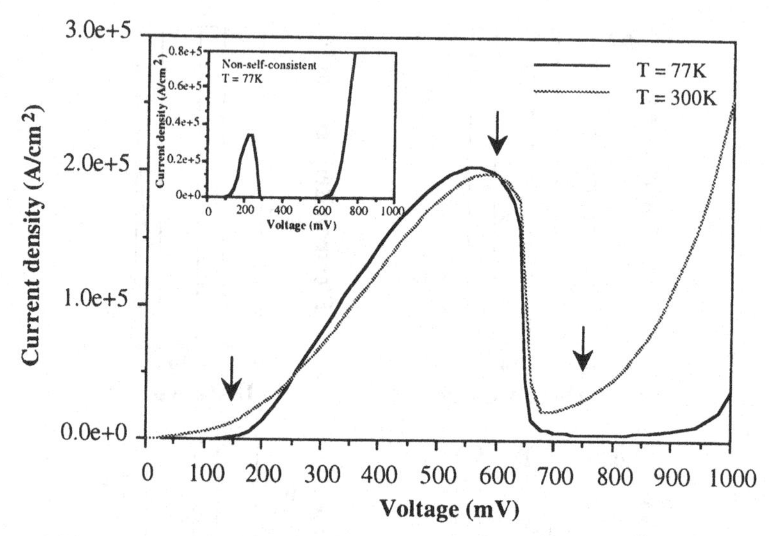

Figure 2.8 *I–V* characteristics calculated self-consistently for Material 1 at 77 K (solid line) and 300 K (grey line). Arrows indicate three typical bias points: threshold, peak and valley voltages. The inset shows the *I–V* curve calculated using the piecewise-linear model.

from the self-consistent calculations is much larger than that from the non-self-consistent calculations, since in the piecewise-linear model of the band diagram the applied voltage falls across the double-barrier region. Calculated conduction-band profiles are also shown for the three typical bias points indicated by arrows in Fig. 2.8: at threshold (Fig. 2.9(a)), at the peak state (Fig. 2.9(b)) and at the valley state (Fig. 2.9(c)).

In self-consistent modelling, a significant potential decrease can be seen in the emitter and collector regions, leading to only a proportion of the applied voltage falling across the double-barrier region. This results in a shift of the current peak towards a larger voltage. The peak voltage obtained from the self-consistent calculations is 0.56 V, which shows good agreement with the value of 0.54 V observed experimentally (see Table 2.3). This demonstrates that self-consistent calculations reproduce the energy-band diagrams in the device very well.

The peak and valley current densities calculated at 77 K and 300 K are also given in Table 2.3, along with the experimental data. Taking into consideration the inevitable uncertainty about layer thickness (due, for instance, to interface roughness), the calculated peak current densities agree well with the experimental data. The calculated peak current is

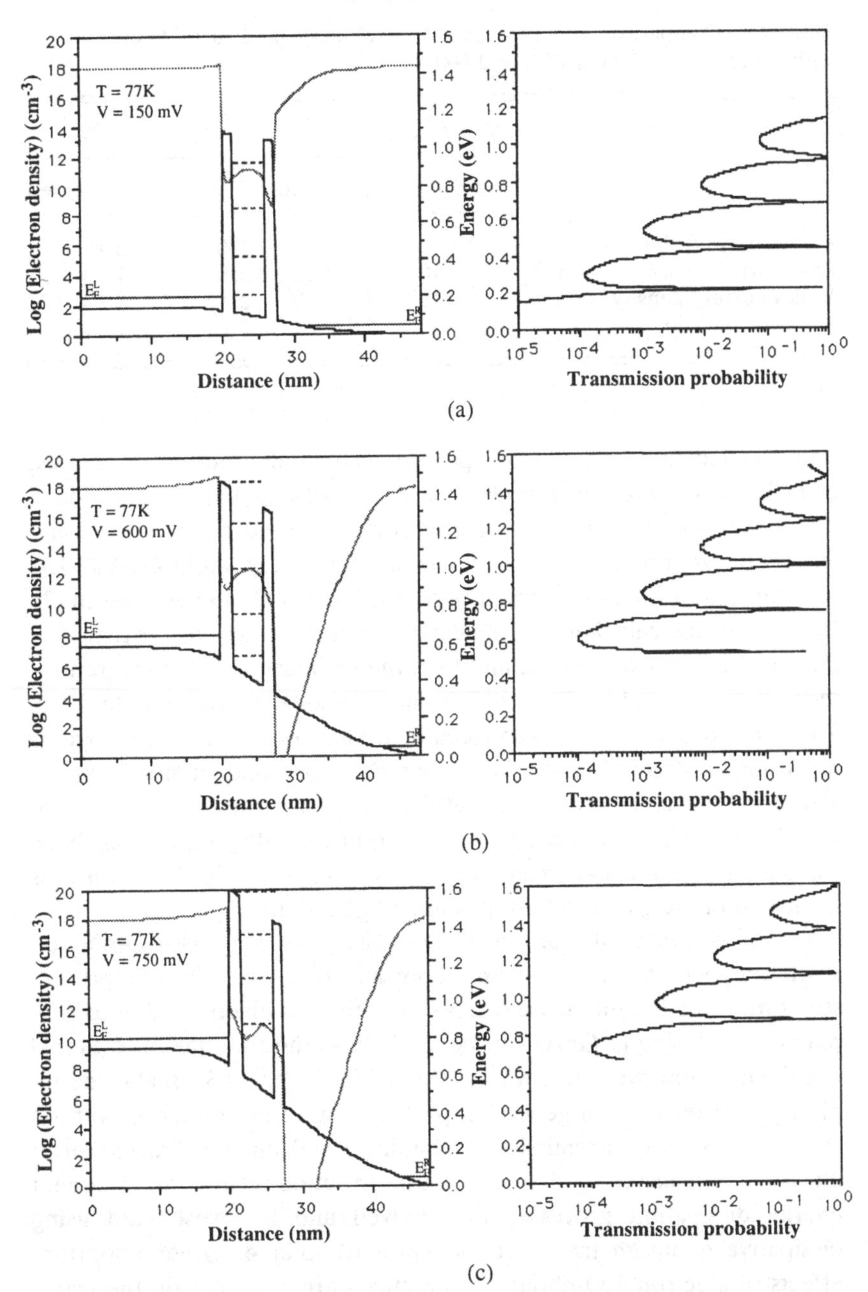

Figure 2.9 Self-consistently calculated energy-band diagrams and transmission properties at three different bias points indicated by the arrows from Fig. 2.8: (a) at current threshold, (b) at the peak current state, and (c) at the valley current state.

Table 2.3. Comparison of the calculated peak voltage and current densities with experimental data at 77 K and 300 K

	$T = 77$ K		$T = 300$ K	
	Calculated	Observed	Calculated	Observed
Peak voltage (mV)	560	540	580	540
Peak current density (A cm^{-2})	2.0×10^5	1.6×10^5	2.0×10^5	1.5×10^5
Valley current density (A cm^{-2})	5.0×10^3	6.2×10^4	2.2×10^4	8.8×10^4
P/*V* current ratio	41	2.6	8.9	1.7

found to decrease by only a few percent as the temperature is increased from 77 K to 300 K, which is also consistent with the experimental data. Good agreement between calculated and measured peak currents has also been reported by Tsuchiya *et al.* for many AlAs/GaAs/AlAs structures with various values of barrier thickness [2] and well width [3]. In contrast, the calculated valley current density is far smaller than that observed in the experiment, and thus the resulting *P*/*V* current ratio is more than one order of magnitude larger than the experimental data. This large difference between calculated and measured valley currents has been similarly observed in various structures and cannot be solely attributed to the uncertainty about the material parameters used in the calculations. The excess current observed in the valley regime has been ascribed to phase-coherence breaking scattering (see Fig. 2.10), which is neglected in the global coherent tunnelling model.

A major cause of the phase-coherence breaking is LO-phonon scattering, which may play an important role even in the present structure since the momentum relaxation time of electrons due to LO-phonon scattering in GaAs can be as short as the time taken to tunnel out the quantum well (this will be studied in Section 2.3). Inelastic LO-phonon scattering is in general expected to have an influence both on the *spectrum* (i.e. transmission probability) and on the *distribution* of electrons. The expected change in the distribution results from frequent energy dissipation processes in the well and is investigated using dissipative quantum transport theory in Chapter 4. Other important effects of electron–LO-phonon interactions are expected on the transmission probability. Firstly, a *real* LO-phonon emission process may open an additional resonant tunnelling channel which can be represented by a satellite transmission peak. Secondly, a *virtual* LO-phonon

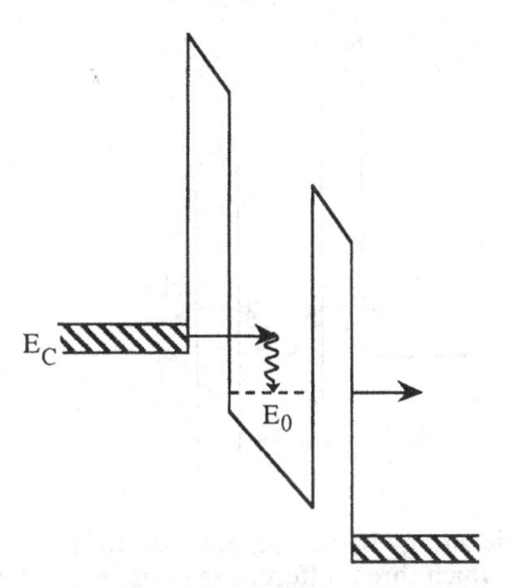

Figure 2.10 Schematic sequential tunnelling process mediated by an energy relaxation process (see Section 2.6), thought to give rise to the discrepancy between measured valley currents and those predicted by the global coherent tunnelling model.

emission–absorption process might cause a broadening of the transmission peak by breaking phase-coherence of electron waves undergoing resonant tunnelling. We will come back to this issue in Chapter 3 and will discuss it in more detail there.

2.2.3 Transfer Hamiltonian model

The transfer matrix theory that we have studied so far is very general and is basically applicable to any resonant tunnelling heterostructure. The theory is suitable for numerical calculations of transmission probability and thus we do not need any other method as long as we are concerned about completely coherent tunnelling. Prior to closing the present section on basic theory, however, it is worth looking at the *transfer Hamiltonian model* since it is a more analytical model than the transfer matrix theory and so is sometimes more suitable when stepping beyond the global coherent tunnelling picture. As a matter of fact, we note that this model is used by many people to analyse the effects of electron–phonon or electron–electron interactions on resonant tunnelling because its extension to the incoherent tunnelling regime is quite straightforward.

In this method the resonant tunnelling structure is divided into three regions (e, w, c), and the electron transfer between these regions is

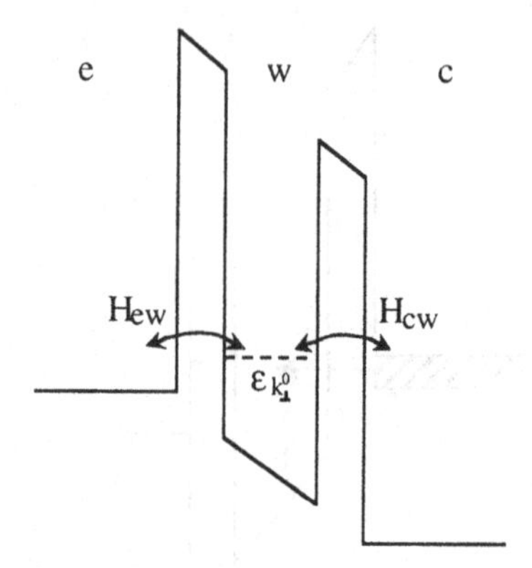

Figure 2.11 A schematic energy-band diagram used for the transfer Hamiltonian model in which three different regions (e, w, c) and tunnelling processes are indicated.

described using tunnelling Hamiltonians through a single barrier (see Fig. 2.11). The total Hamiltonian of the system, H^{total}, is expressed as follows:

$$H^{\text{total}} = H_{\text{e}}^{0} + H_{\text{c}}^{0} + H_{\text{w}} \tag{2.29}$$

$$H_{\text{w}} = H_{\text{w}}^{0} + H_{\text{ew}} + H_{\text{cw}} + H_{\text{other}} \tag{2.30}$$

In these equations H_{e}^{0}, H_{c}^{0} and H_{w}^{0} are the Hamiltonians of free electrons in the emitter, collector and quantum well,

$$H_{\text{e}}^{0} = \sum_{\mathbf{k}_{/\!/}^{\text{e}}, k_{\perp}^{\text{e}}} (\epsilon_{\mathbf{k}_{/\!/}^{\text{e}}} + \epsilon_{k_{\perp}^{\text{e}}}) \overset{+}{a}_{\mathbf{k}_{/\!/}^{\text{e}}, k_{\perp}^{\text{e}}} a_{\mathbf{k}_{/\!/}^{\text{e}}, k_{\perp}^{\text{e}}} \tag{2.31}$$

$$H_{\text{c}}^{0} = \sum_{\mathbf{k}_{/\!/}^{\text{c}}, k_{\perp}^{\text{c}}} (\epsilon_{\mathbf{k}_{/\!/}^{\text{c}}} + \epsilon_{k_{\perp}^{\text{c}}}) \overset{+}{a}_{\mathbf{k}_{/\!/}^{\text{c}}, k_{\perp}^{\text{c}}} a_{\mathbf{k}_{/\!/}^{\text{c}}, k_{\perp}^{\text{c}}} \tag{2.32}$$

$$H_{\text{w}}^{0} = \sum_{\mathbf{k}_{/\!/}^{\text{w}}} (\epsilon_{\mathbf{k}_{/\!/}^{\text{w}}} + \epsilon_{k_{\perp}^{0}}) \overset{+}{a}_{\mathbf{k}_{/\!/}^{\text{w}}, k_{\perp}^{0}} a_{\mathbf{k}_{/\!/}^{\text{w}}, k_{\perp}^{0}} \tag{2.33}$$

where $\mathbf{k}_{/\!/}$ and $k_{\perp}$ denote the lateral and vertical components of an electron wavevector, $\mathbf{K} = (\mathbf{k}_{/\!/}, k_{\perp})$, $\epsilon_{k_{\perp}^{0}}$ is the energy of the resonant state $(=E_0)$ and $\overset{+}{a}_{\mathbf{k}}$ and $a_{\mathbf{k}}$ are creation and annihilation operators for electrons. In eqn (2.30) H_{w} is the total Hamiltonian of electrons in the quantum well which includes tunnelling terms between the emitter (collector) and quantum well, $H_{\text{ew}}(H_{\text{cw}})$,

$$H_{ew} = \sum_{\mathbf{k}^e_{\parallel},k^e_{\perp},\mathbf{k}^w_{\parallel}} \{V(\mathbf{k}^e_{\parallel},k^e_{\perp};\mathbf{k}^w_{\parallel},k^0_{\perp}) a^+_{\mathbf{k}^e_{\parallel},k^e_{\perp}} a_{\mathbf{k}^w_{\parallel},k^0_{\perp}} + H.c.\} \tag{2.34}$$

$$H_{cw} = \sum_{\mathbf{k}^c_{\parallel},k^c_{\perp},\mathbf{k}^w_{\parallel}} \{V(\mathbf{k}^c_{\parallel},k^c_{\perp};\mathbf{k}^w_{\parallel},k^0_{\perp}) a^+_{\mathbf{k}^c_{\parallel},k^c_{\perp}} a_{\mathbf{k}^w_{\parallel},k^0_{\perp}} + H.c.\} \tag{2.35}$$

where $V(\mathbf{k}^e_{\parallel},k^e_{\perp};\mathbf{k}^w_{\parallel},k^0_{\perp})$ and $V(\mathbf{k}^c_{\parallel},k^c_{\perp};\mathbf{k}^w_{\parallel},k^0_{\perp})$ are the tunnelling matrix elements between the reservoirs and quantum well, and the last term, H_{other}, represents other interactions which the electrons may experience in the well (see, for example, Section 3.1 or Section 3.3).

In this formalism, the transmission probability may be expressed by using the Green's function of the electrons in the well, G_w, as follows:

$$\begin{aligned} T(\mathbf{k}^e_{\parallel},k^e_{\perp}) &= -\sum_{\mathbf{k}^c_{\parallel},k^c_{\perp}} \mathrm{Im}\langle \mathbf{k}^c_{\parallel},k^c_{\perp}|H_{cw}G_wH_{ew}|\mathbf{k}^e_{\parallel},k^e_{\perp}\rangle \\ &= -\sum_{\mathbf{k}^c_{\parallel},k^c_{\perp}} \mathrm{Im}\left\langle \mathbf{k}^c_{\parallel},k^c_{\perp}\left| H_{cw}\frac{1}{E(\mathbf{k}^e_{\parallel},k^e_{\perp}) - i\delta - H_w}H_{ew}\right| \mathbf{k}^e_{\parallel},k^e_{\perp}\right\rangle \\ &= -\sum_{\mathbf{k}^c_{\parallel},k^c_{\perp}\mathbf{k}^w_{\parallel}}\sum \mathrm{Im}\frac{\langle \mathbf{k}^c_{\parallel},k^c_{\perp}|H_{cw}|\mathbf{k}^w_{\parallel},k^0_{\perp}\rangle\langle \mathbf{k}^w_{\parallel},k^0_{\perp}|H_{ew}|\mathbf{k}^e_{\parallel},k^e_{\perp}\rangle}{E(\mathbf{k}^e_{\parallel},k^e_{\perp}) - E(\mathbf{k}^w_{\parallel},k^0_{\perp}) - \Sigma_w(\mathbf{k}^w_{\parallel},k^0_{\perp},E(\mathbf{k}^e_{\parallel},k^e_{\perp}))} \end{aligned} \tag{2.36}$$

where Σ_w is the self-energy of electrons in the well due to both the tunnelling and other interaction terms. Since the tunnelling Hamiltonians, H_{ew} and H_{cw}, conserve momentum, $V(\mathbf{k}_{\parallel},k_{\perp};\mathbf{k}^w_{\parallel},k^0_{\perp}) = V_{k_{\perp}}$, and

$$T(\mathbf{k}^e_{\parallel},k^e_{\perp}) = -\mathrm{Im}\frac{\Gamma}{E(\mathbf{k}^e_{\parallel},k^e_{\perp}) - E(\mathbf{k}^e_{\parallel},k^0_{\perp}) - \Sigma_w(\mathbf{k}^e_{\parallel},k^0_{\perp},E(\mathbf{k}^e_{\parallel},k^e_{\perp}))} \tag{2.37}$$

where Γ is the intrinsic broadening of the resonant state given by

$$\Gamma \approx \sum_{k_{\perp}}|V_{k_{\perp}}|^2\delta(\epsilon_{k_{\perp}} - \epsilon_{k^0_{\perp}}) \tag{2.38}$$

When the electrons have no interaction in the well ($H_{other} = 0$), the lowest-order term of the transmission probability, $T^{(0)}(\mathbf{k}^e_{\parallel},k^e_{\perp})$, reduces to the following simple expression:

$$T^{(0)}(\mathbf{k}^e_{\parallel},k^e_{\perp}) = T^{(0)}(\epsilon_{k^e_{\perp}}) = \frac{\Gamma^2}{(\epsilon_{k^e_{\perp}} - \epsilon_{k^0_{\perp}})^2 + \Gamma^2} \tag{2.39}$$

by replacing the self-energy, Σ_w, in eqn (2.37) by $-i\Gamma$. This is in fact the Breit–Wigner formula which we have previously derived within the transfer matrix theory (see eqn (2.23)). The total tunnelling current through the double barriers is then expressed by:

$$J = \frac{em^*}{2\pi\hbar^3}\int_0^\infty d\epsilon_{\mathbf{k}_{/\!/}}\int_0^\infty d\epsilon_{k_\perp}(f_L(\epsilon_{\mathbf{k}_{/\!/}},\epsilon_{k_\perp}) - f_R(\epsilon_{\mathbf{k}_{/\!/}},\epsilon_{k_\perp}))T(\mathbf{k}_{/\!/},k_\perp) \quad (2.40)$$

which gives the same tunnelling current as that calculated by the global coherent tunnelling theory, provided there are no scattering events in the quantum well. A great advantage of this formula is that the effects of scattering processes can be introduced naturally through the self-energy term in eqn (2.36), and thus this is sometimes used to analyse resonant tunnelling phenomena mediated by various scattering processes (this will be studied in Chapter 3).

2.3 Electron dwell time

This section describes the characteristic time of RTDs, the electron dwell time, τ_d, which is very important in high-speed applications. The dwell time is the time which is required for an electron to tunnel through the double barriers, or, equivalently, the time for which an electron stays between the double barriers. In a classical fluid-dynamic picture the dwell time is defined by using the following simple current continuity equation:

$$\tau_d = \frac{e \cdot \sigma_w}{J} \quad (2.41)$$

where J is the tunnelling current density and σ_w is the sheet concentration of electrons in the double-barrier structure. Equation (2.41) means that the dwell time can be determined by using J and σ_w, which can be measured experimentally.

In a quantum mechanical picture the dwell time is defined as the tunnelling delay time of the incident electron wave, which is obtained via the steady-state phase shift, δ, as follows [14]:

$$\tau_d = \hbar\frac{\partial\delta(E_z)}{\partial E_z} \quad (2.42)$$

The phase shift, δ, is defined by the following expression:

$$\frac{\Psi_R}{\Psi_L} = \left|\frac{\Psi_R}{\Psi_L}\right|\exp(i\delta) \quad (2.43)$$

By using the Breit–Wigner formula (2.23), we find that

$$\delta = \delta_0 - \tan^{-1}\left(\frac{\Gamma}{E_z - E_z^0}\right) \quad (2.44)$$

$$\delta_0 = -k_z(2d + D) \tag{2.45}$$

From eqn (2.42) and (2.44) we have

$$\tau_d(E_z) = \hbar \frac{\Gamma}{(E_z - E_z^0)^2 + \Gamma^2} \tag{2.46}$$

which at resonance leads to

$$\tau_d = \hbar/\Gamma \tag{2.47}$$

If we use the approximate expression (2.25) for Γ, we finally obtain the following expression for the dwell time:

$$\tau_d \approx \frac{1}{2T_{SB}(E_z^0)} \frac{D + 2/\kappa_z^0}{v_z^0} \tag{2.48}$$

This proves to be equivalent to the classical *multiple reflection time* of an electron moving in a quantum well with velocity v_z^0 (see Fig. 2.12). The term, $2/\kappa_z^0$, in eqn (2.48) is a quantum mechanical correction to the well width owing to the penetration of an electron into the potential barriers.

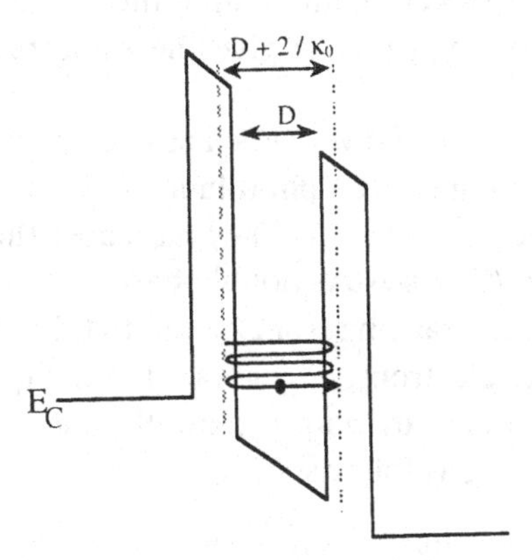

Figure 2.12 Schematic energy-band diagram of a double-barrier RTD and the multiple-reflection process of an electron.

Numerical calculations of the time evolution of a wave packet have been performed by Harada and Kuroda [15] in order to evaluate the dwell time. Assuming that the electrostatic potential does not change with time, the time evolution of the wave packet may be given by the following equation:

$$\Psi(z,t) = \frac{1}{\sqrt{(\Delta k_z)}} \int \Psi(z,k_z) \exp\left(-\frac{(k_z - k_z^0)^2}{2\Delta k_z^2} - i\frac{E(k_z)}{\hbar} t \right) dk_z \quad (2.49)$$

where $\Psi(z,k_z)$ is a steady scattering state with wavenumber, k_z, obtained by solving the time-independent Schrödinger equation, eqn (2.9). In eqn (2.49) the electron distribution is assumed to be Gaussian in k_z-space, centred at k_z^0 and of width Δk_z. The transient behaviour of the wave packet is shown in Fig. 2.13(a)–(f) for a double-barrier structure with a barrier thickness of 2 nm, barrier height of 0.4 eV and well width of 5 nm.

The central wavenumber, k_z^0, is chosen so that the energy of the incident wave coincides with the energy of the resonant state. The amplitude of the wavefunction in the quantum well decreases exponentially and thus the dwell time can be obtained as the time-constant of the decay. The calculated dwell times are plotted in Fig. 2.14 as a function of barrier thickness for various values of well width (solid circles). Also shown as solid lines are the dwell times obtained using eqn (2.47) and the width of the resonant state, Γ (see the right-hand axis of the diagram), obtained from the transmission probability function. Overall agreement between these results indicates that the calculations based on the transient wave-packet method and those based on the simple uncertainty principle (eqn (2.47)) give the same results for the dwell time.

An experimental study of dwell time has been performed by Tsuchiya *et al.* [16] using a time-resolved photoluminescence (PL) measurement technique with a picosecond laser. They measured the time evolution of PL spectra for AlAs/GaAs/AlAs double-barrier structures with various barrier thicknesses, L_B, ranging from 2.8 nm to 6.2 nm. The decay of the concentration of the electrons, n, generated in the quantum well by the laser pulse is determined by both the dwell time, τ_d, and the radiative recombination time, τ_r, as follows:

$$\frac{dn}{dt} = -\frac{n}{\tau_d} - \frac{n}{\tau_r} \quad (2.50)$$

The extracted PL decay times at various temperatures are shown in Fig. 2.15 as a function of barrier thickness. The dot–dashed line represents the electron dwell time calculated from eqn (2.47). Another theoretical result assuming a higher-energy barrier height of 1.36 eV (Dingle's rule) is also indicated by the broken line. It is found that the measured decay time is in agreement with the theoretical values of τ_d in

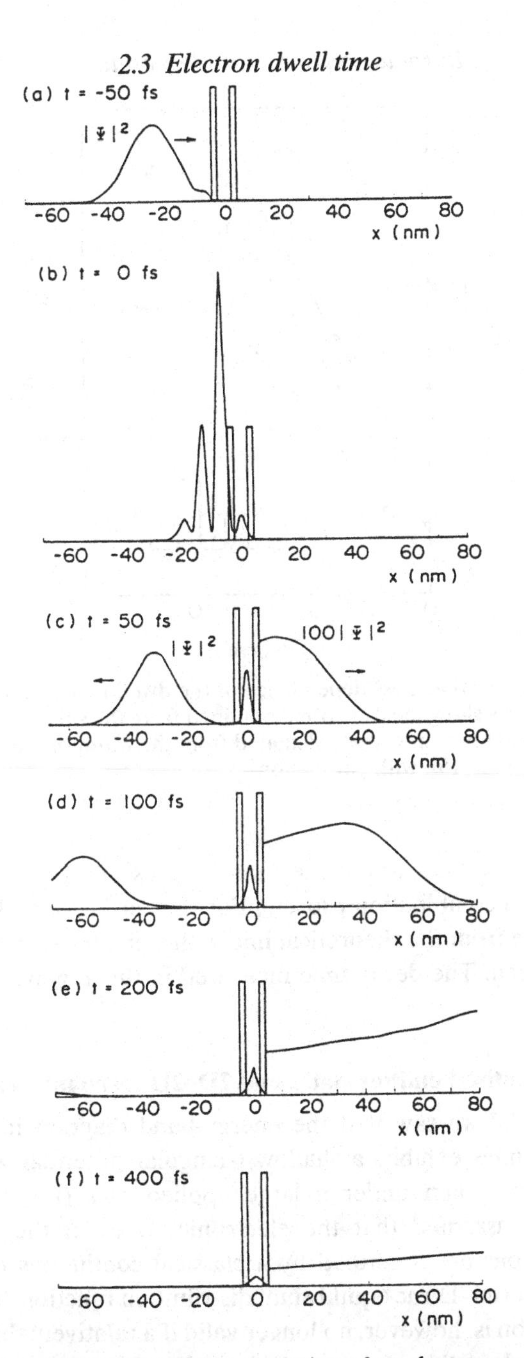

Figure 2.13 Potential profile and time evolution of an electron wave packet in a resonant tunnelling system with barrier height V of 400 mV, barrier thickness d of 2 nm, well width L of 5 nm, average wavenumber of the incident wave packet k_0 of 4.44×10^8 m^{-1} and wavenumber width Δk of 1×10^8 m^{-1}. After Harada *et al.* [15], with permission.

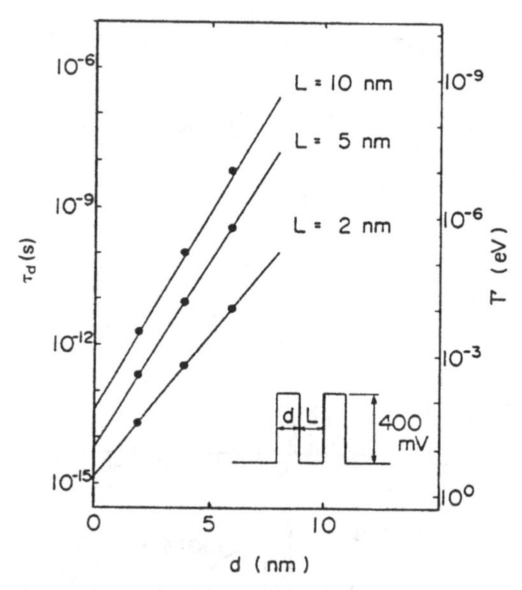

Figure 2.14 Barrier thickness dependence of the dwell time at the first resonant state. Solid circles show the dwell time obtained from the simulation of the wave packet, and solid lines show that calculated from the width of the resonant level. After Harada *et al.* [15], with permission.

the regime $L_B < 4$ nm where the tunnelling escape process dominates the radiative recombination process. For thicker barriers the measured values deviate from the theoretical line with some temperature dependence being seen. The decay time measured in this regime is thought to reflect τ_r.

2.4 Quantised emitter states and 2D–2D resonant tunnelling

In Section 2.2.2 we saw that the energy-band diagram in the emitter region sometimes exhibits a shallow triangular potential well with the emitter barrier when under a large applied bias (for example, see Fig. 2.9). We assumed that the electronic states in the emitter and collector regions are described by a classical continuous energy spectrum with a Fermi–Dirac equilibrium distribution function. This convenient assumption is, however, no longer valid if a relatively thick undoped (or even low-doped) spacer layer is introduced on the emitter side. In such cases, the external voltage falls across these undoped regions as well as across the double-barrier structure, resulting in a pseudo-triangular quantum well, as shown in Fig. 2.16. In general, a low-doping

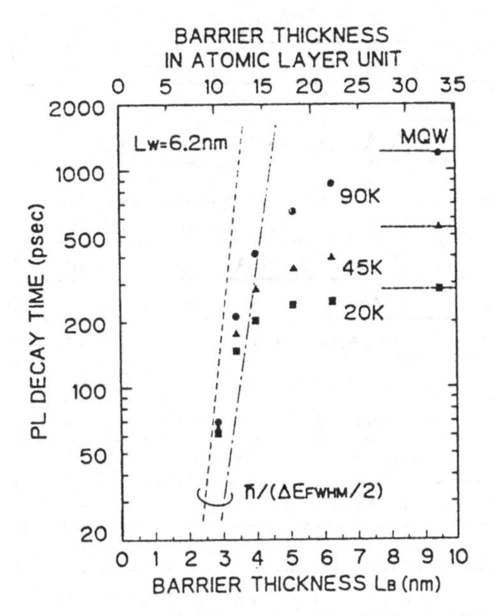

Figure 2.15 PL decay times plotted as functions of barrier thickness L_B. When $L_B < 4$ nm, the tunnelling escape process is dominant. The broken and dot–dashed lines are theoretical dwell times calculated assuming barrier heights of 1.36 eV and 0.96 eV respectively. After Tsuchiya *et al.* [16], with permission.

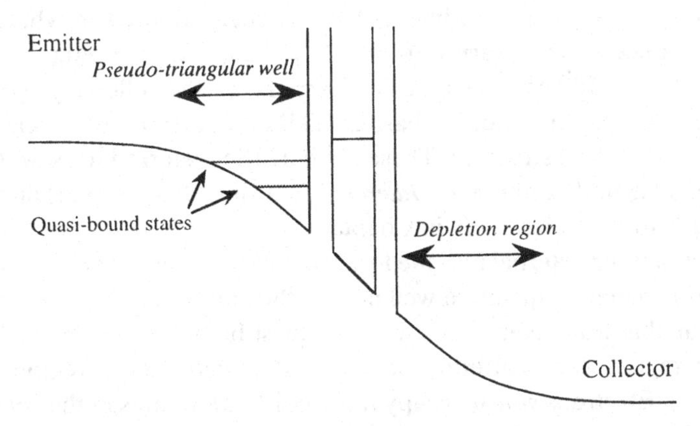

Figure 2.16 Schematic energy-band diagram of a double-barrier RTD with thick low-doped regions on both side of the barriers.

profile near to the resonant tunnelling barriers gives a larger P/V current ratio and hence is widely used, even though the resulting current densities are lower.

An example of an advanced contact structure is shown in Fig. 2.17; in

Emitter

Layer	Thickness
n-GaAs (Si: 5 x 10^{18}/cm^3)	150 nm
n-GaAs (Si: 2 x 10^{17}/cm^3)	150 nm
n-GaAs (Si: 8 x 10^{16}/cm^3)	200 nm
n-GaAs (Si: 3 x 10^{16}/cm^3)	250 nm
n-GaAs (Si: 1 x 10^{16}/cm^3)	450 nm
undoped-GaAs	5 nm
Double barrier structure	
undoped-GaAs	5 nm
n-GaAs (Si: 1 x 10^{16}/cm^3)	450 nm
n-GaAs (Si: 3 x 10^{16}/cm^3)	250 nm
n-GaAs (Si: 8 x 10^{16}/cm^3)	200 nm
n-GaAs (Si: 2 x 10^{17}/cm^3)	150 nm
n-GaAs (Si: 5 x 10^{18}/cm^3)	500 nm

Collector

Figure 2.17 Emitter and collector structures with a graded doping profile used for Materials 2 and 3.

this structure a graded doping profile has been adopted in which the doping concentration varies from $5 \times 10^{18}\ \mathrm{cm}^{-3}$ at the edge of the emitter to $1 \times 10^{16}\ \mathrm{cm}^{-3}$ near to the barriers. In the following sections two AlAs/GaAs/AlAs double-barrier RTDs are introduced which have this graded contact structure. These are hereafter referred to as *Material 2* (4.2 nm/5.9 nm/4.2 nm) and *Material 3* (5.0 nm/7.0 nm/5.0 nm) and will be frequently used for study in Chapters 4 and 6.

Under a large external bias the low-doping region in the emitter forms a pseudo-triangular quantum well next to the emitter barrier. Electronic states in this triangular well are both quasi-bound states in the low-energy region and scattering states in the high-energy region (see Fig. 2.16). Electrons which occupy the quasi-bound states in the triangular well form a 2D electron gas (2D EG). Resonant tunnelling then occurs between these 2D states in the triangular well and the resonant state in the double-barrier structure: this is hence termed *2D–2D resonant tunnelling* [17] and is distinguished from conventional resonant tunnelling, which should now be denoted 3D–2D resonant tunnelling in a corresponding manner.

The 2D–2D resonant tunnelling characteristics are different from the

3D–2D ones since both electron systems in the emitter and well now have discrete energy levels. This results in a sharper resonance when the lowest energy level in the well is aligned with the lowest level in the emitter. Figure 2.18(a) and (b) [18] show typical characteristics for RTDs fabricated on Material 2 and Material 3 respectively. The *I–V* curves are shown in one bias direction only as the characteristics in the

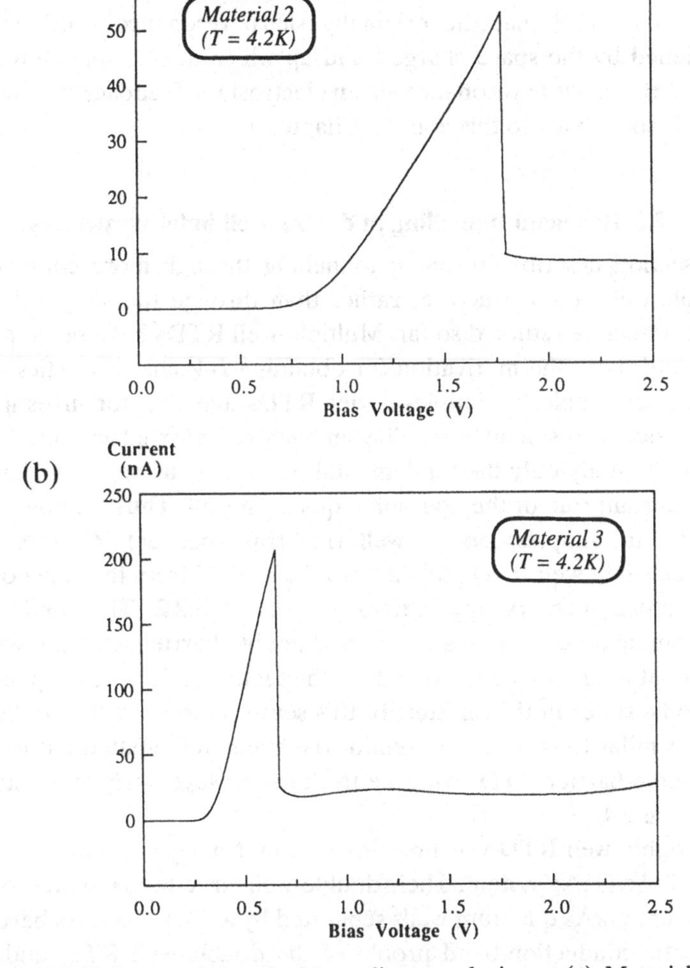

Figure 2.18 *I–V* characteristics of a 14 μm diameter device on (a) Material 2 and (b) Material 3 measured at 4.2 K. The characteristics are shown on the same voltage scale, but the current scales differ by a factor of 230. After Goodings [18], with permission.

reversed direction are essentially the same, with only slight differences in the currents and voltages due to inevitable asymmetry in the double barriers. Material 2 shows a higher peak current and peak voltage than does Material 3, as is expected from the differences in barrier and well thickness. Both materials show relatively sharp current peaks with large *P/V* current ratios which can be taken as evidence of 2D–2D resonant tunnelling. The present issue is closely related to the non-equilibrium distribution and space charge build-up in the emitter and well and is worthy of more consideration; it has indeed been demonstrated by Alves *et al.* [19] that the originally sharp resonance width can be broadened by the space charge build-up which acts to maintain alignment of the levels at resonance via an electrostatic feedback mechanism. We will come back to this issue in Chapter 4.

2.5 Resonant tunnelling in double-well heterostructures

This section describes resonant tunnelling through more complicated multiple-well heterostructures rather than through the simple double-barrier structures studied so far. Multiple-well RTDs have been investigated with both the motivation for obtaining *I–V* characteristics which are not achievable by double-barrier RTDs and that for investigating the physics of resonant tunnelling in such complex structures. In this section we study only the fundamental physics, focusing particularly on the important role of the additional quantum well. Thus we now take a look at the simplest double-well (i.e. triple-barrier) RTD. A more complex triple-well RTD [20]–[22] will be studied from the viewpoint of multi-valued memory applications in Section 5.2.2. The double-well RTD can be thought of as an improved double-barrier structure with an additional quantum well inserted on the emitter side to give quantised electronic states in the emitter. In this sense the role of this additional well is similar to that of the pseudo-triangular well created naturally in the double-barrier RTDs with the thick spacer layers which we studied in Section 2.4.

A double-well RTD was first reported by Nakagawa *et al.* [23] using an AlGaAs/GaAs system. Their double-well structure consisted of two equivalent GaAs quantum wells separated by a thick AlGaAs barrier; a schematic conduction-band profile of the double-well RTD under an applied bias is shown in Fig. 2.19. In this structure electrons are first injected from the 3D emitter to the lowest 2D states in well 1 and can then tunnel to the quasi-bound states in well 2 when two energy levels

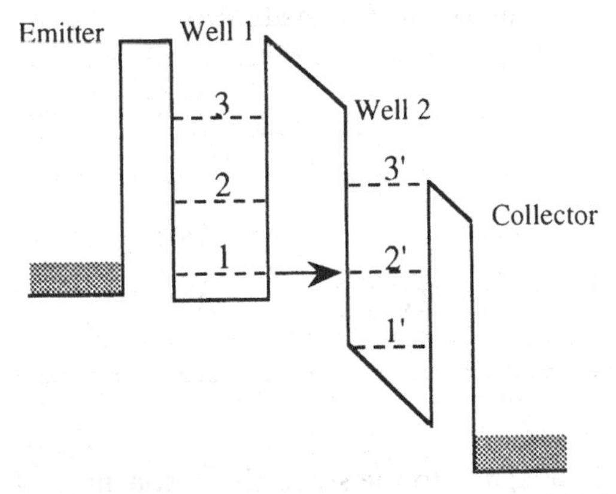

Figure 2.19 Schematic conduction-band profile of a double-well RTD at resonance 1 → 2'. After Nakagawa *et al.* [23], with permission.

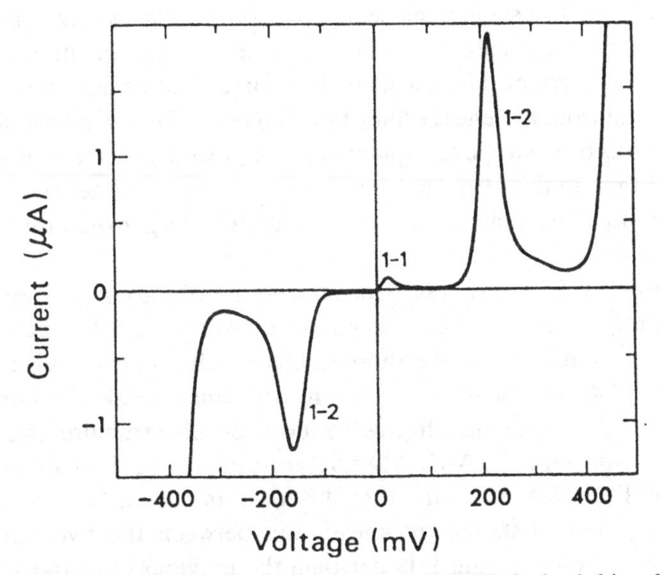

Figure 2.20 *I–V* characteristics for a double-well RTD in both bias directions measured at 4.2 K. After Nakagawa *et al.* [23], with permission.

are aligned. Since the thick barrier between well 1 and well 2 enhances the electron accumulation in well 1, the conduction-band profile is expected to be piecewise-linear, as illustrated in Fig. 2.19; this has in fact been proved to be a good approximation through self-consistent numerical calculations of the band profile.

A typical *I–V* characteristic observed by Nakagawa *et al.* for this device is shown in Fig. 2.20. The multiple current peaks observed have

Table 2.4. A layer structure of the InGaAs/InAlAs double-well RTD

Layer		Thickness (nm)
i-$In_{0.53}Ga_{0.47}As$	L_{SPACER}	10.0
i-$In_{0.52}Al_{0.48}As$	L_{B1}	2.5
i-$In_{0.53}Ga_{0.47}As$	L_{W1}	5.0
i-$In_{0.52}Al_{0.48}As$	L_{B2}	5.0
i-$In_{0.53}Ga_{0.47}As$	L_{W2}	15.0
i-$In_{0.52}Al_{0.48}As$	L_{B3}	2.5

been successfully assigned to the successive resonances of quasi-bound states of well 2 (denoted by numbers 1', 2', 3', . . .), with the lowest quasi-bound state of well 1 (denoted by 1); this state is hereafter termed an injection level. The small peak seen close to zero bias is attributable to $1 \rightarrow 1'$ resonant tunnelling, which occurs due to the inevitable asymmetry of the barriers. The excellent *P*/*V* current ratios achieved in this device result from the energy filter function of the injection level for the incoming electron waves, as explained in the preceding section. It should be noted that, in this structure, peak currents due to higher resonances are generally more than one order of magnitude larger than that at the first resonance.

Triple-well RTDs have also been studied in order to investigate their use in multiple-valued logic applications. Details are left for their study in Chapter 5, but one crucial requirement for such applications is to have multiple NDC characteristics with nearly equal peak currents. To investigate this issue an alternative double-well structure has been proposed using an InGaAs/InAlAs heterosystem. The layer structure is shown in Table 2.4. This structure differs from the earlier one in the following points: first, the thickness ratio between the two quantum wells is three, which is much larger than the previous one; and second, the thinner quantum well (denoted W1) is placed on the emitter side and used as an injection well (the wider well is denoted W2 in the same way as in the preceding section). The principle of operation and the advantages of this structure in realising high peak current density and uniform NDC peaks are described below.

The large thickness ratio between the two quantum wells leads to a large difference between the lowest quasi-eigenenergies of the wells so that the resonances are expected to take place between the ground state of W1 and the highly excited states of W2. The energy-band diagram

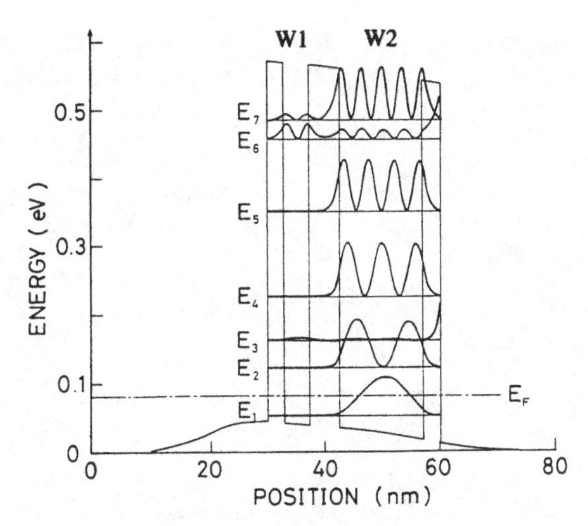

Figure 2.21 Energy-band diagram and associated quasi-eigenstates of an InGaAs/InAlAs double-well RTD calculated at zero bias.

and associated quasi-eigenstates are shown in Fig. 2.21. Since the quasi-eigenstates displayed in this diagram are those calculated for an incident wave from the collector side, it may be a little difficult to identify the quasi-eigenstates of W1; the states indicated by E_3 and E_6 are the lowest and first excited quasi-bound states of W1. As seen in this diagram, resonances of the second, third and fourth excited states of W2 (E_4, E_5 and E_7) with the injection level E_3 lead to multiple NDCs in the *I–V* characteristics. The use of the excited states provides high peak current densities as well as good *P/V* ratios which are attributed to less scattering events in the present InGaAs/InAlAs system than in the conventional AlGaAs/GaAs systems.

The *I–V* characteristics for a $6 \times 6\ \mu m^2$ diode measured at 127 K and 179 K are shown in Fig. 2.22(a) and (b). The differences between the peak currents are found to be much smaller than those for the AlGaAs/GaAs double-well RTD, as expected. The *I–V* characteristics calculated using the self-consistent global coherent tunnelling model (Section 2.2.2) are shown in Fig. 2.22(c), along with the experimental data. (The material parameters of the InP lattice-matched InGaAs/InAlAs hetero-system are listed in Table 2.5.) The overall agreement between the measured and calculated peak voltages is again very good. It should be pointed out again that the excellent agreement between all the calculated and observed peak voltages confirms that self-consistent modelling can reproduce the potential distribution very well. Relatively good

(a)

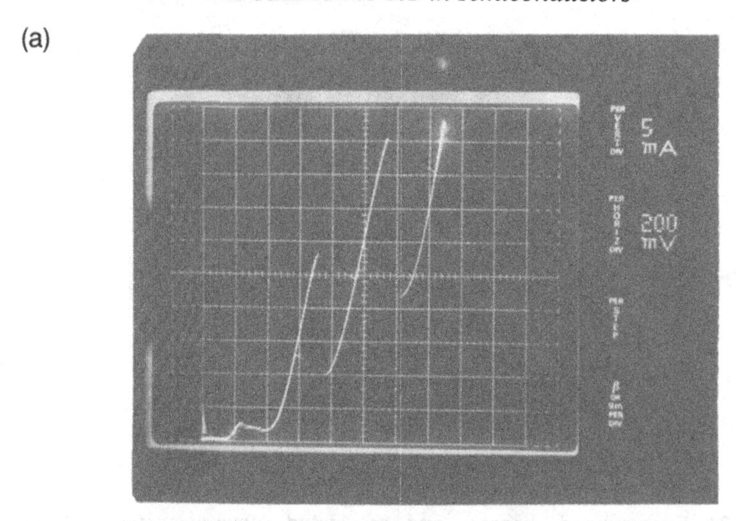

(b)

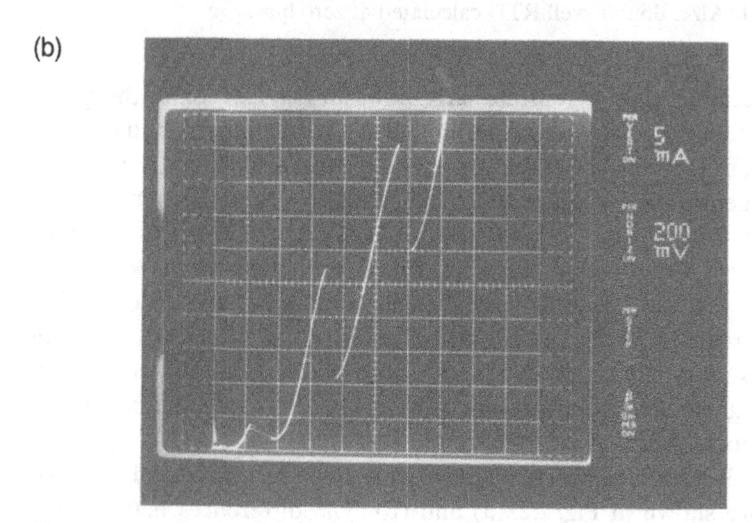

Figure 2.22 *I–V* characteristics of an InAlAs/InGaAs double-well RTD with a mesa area of 6 μm × 6 μm measured at (a) 127 K and (b) 179 K. Part (c) shows the *I–V* curve calculated self-consistently at 77 K, along with the experimental data at 84 K.

agreement has also been seen for peak currents, but the observed current peaks have turned out to be much broader than the calculated ones, as we have previously seen for double-barrier RTDs. Consequently, the observed valley currents are higher than those given by the calculations. From the temperature dependence of the valley currents,

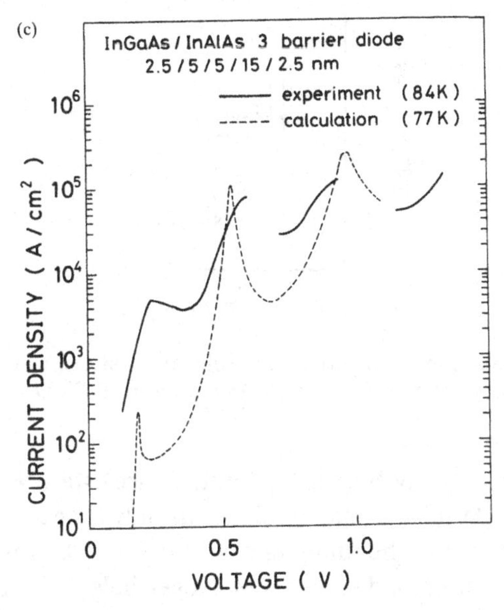

Figure 2.22 *cont.*

we would expect that LO-phonon scattering contributes at least partly to the broadening of the peaks, as mentioned earlier.

It is worth noting that, at the third resonance, the top of the third barrier adjacent to the collector falls below the resonant level E_7, as shown in Fig. 2.23. The third resonance is not a resonance between two quasi-bound states like the first or second resonances. Instead, it is a

Table 2.5. Material parameters of InP lattice-matched InGaAs/InAlAs heterostructures

1. *Electron effective mass* m^*
$m^*(In_{0.53}Ga_{0.47}As) = 0.044m_0$
$m^*(In_{0.52}Al_{0.48}As) = 0.084m_0$

2. *Conduction band discontinuity* ΔE_c
$\Delta E_c = \beta(E_g^\Gamma(In_{0.52}Al_{0.48}As) - E_g^\Gamma(In_{0.53}Ga_{0.47}As))$
$E_g^\Gamma(In_{0.53}Ga_{0.47}As) = 0.71$ eV
$E_g^\Gamma(In_{0.52}Al_{0.48}As) = 1.51$ eV
Band parameter $\beta = 0.667$

3. *Dielectric constant* ϵ
$\epsilon(In_{0.53}Ga_{0.47}As) = 13.9$
$\epsilon(In_{0.52}Al_{0.48}As) = 12.45$

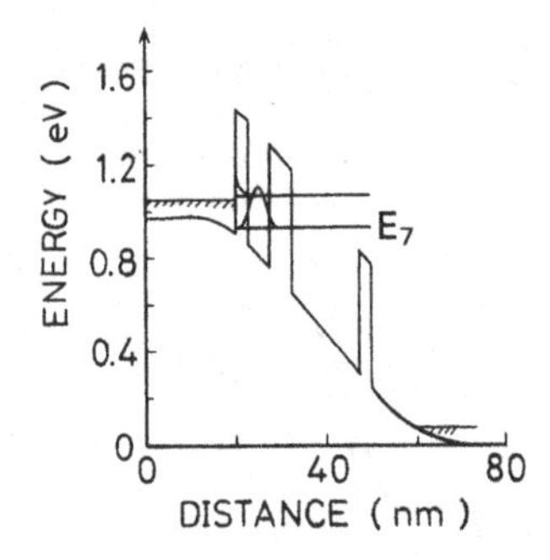

Figure 2.23 Conduction-band profile and relevant scattering state calculated at the third resonance of the InGaAs/InAlAs double-well RTD.

resonance of a virtually bound (or interference) state E_7 with a quasi-bound state E_3. At this resonance, electrons only tunnel through the first two barriers and, thus, the tunnelling current could be much larger than those at the first or second resonances. Nevertheless, the peak current is just twice as large as the second current. This may be attributable to the use of an injection level; the injection level functions as an energy filter, as discussed before, and limits the total number of electrons tunnelling through the structure.

2.6 Phase-coherence breaking scattering and sequential tunnelling

We have studied the global coherent tunnelling model, in which it is assumed that the phase-coherence of the electron waves is completely maintained. The electrons involved in the resonant tunnelling process stay in the double-barrier structure for a dwell time ranging from a few tens of femtoseconds to a few hundred nanoseconds depending on the structural parameters. While in the structure, the electrons may suffer from scattering processes which cause momentum and energy relaxations which break the phase-coherence. As mentioned earlier, several possible causes exist for this sort of phase-coherence breaking scattering: electron–phonon interactions, residual impurity scattering, disorder scattering, and so on. A question then arises as to the use of the global coherent tunnelling model: to what extent is this model valid under these circumstances? The typical mean free time of electrons in a bulk GaAs material is of the order of subpicoseconds at room temperature, which can be much shorter than the dwell time of relatively thick double-barrier structures. For instance, the momentum relaxation time, τ_m, of electrons in a low-doped n-type GaAs bulk material

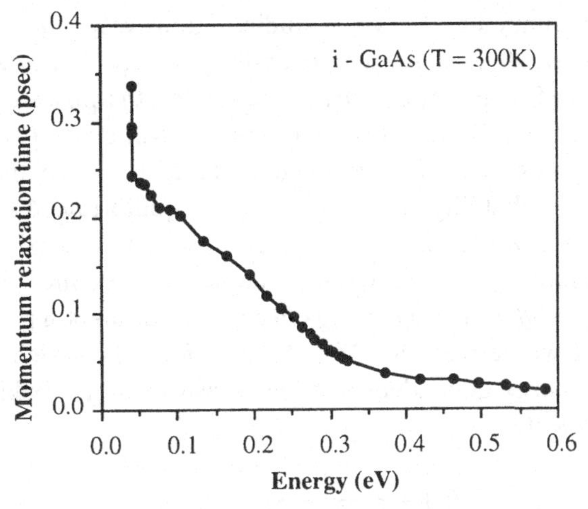

Figure 2.24 Momentum relaxation time of electrons as a function of energy calculated for n^--GaAs at room temperature by a Monte Carlo simulation.

($N_D = 10^{14}$ cm^{-3}) is of the order of 0.1 ps for a wide range of energy, as shown in Fig. 2.24, which is calculated at room temperature using a conventional Monte Carlo simulation which takes account of scattering due to phonons and ionised impurities. It is thus likely that the electrons in the quantum well experience some degree of scattering during the tunnelling process and lose their phase-coherence.

A *sequential tunnelling model* was proposed by Luryi [24], [25], based on the above considerations as an alternative explanation of the RTD phenomenon, about ten years after global coherent resonant tunnelling was reported by Tsu and Esaki. This model describes resonant tunnelling as two continuous tunnelling processes, that is, tunnelling from the emitter into the quantum well followed by that to the collector, as is assumed by the transfer Hamiltonian model (Section 2.2.3). Between these two processes electrons suffer from frequent phase-coherence breaking scattering processes in the quantum well and are well relaxed into locally quasi-equilibrium states. The sequential tunnelling model is thus defined as a picture of the phenomenon at the incoherent extreme. It should be noted that, in the limit of no scattering in the well, the sequential tunnelling model naturally reduces to the coherent global tunnelling model, as is expected from the definition of the model; this has been demonstrated theoretically by Weil and Vinter [26]. Throughout this book, therefore, we will use the term 'sequential tunnelling' for the process in which phase-coherence breaking scattering is dominant.

Sequential tunnelling has been studied intensively both theoretically and experimentally, as its understanding is very important for the development of quantum devices which rely on the phase-coherence of electrons. In the following two chapters, we will study two important physical aspects of sequential tunnelling: the effects of scattering on the transmission probability function and non-equilibrium distribution of electrons are investigated in Chapters 3 and 4 respectively. In this section a simple phenomenological model is presented in order to illustrate the *scattering broadening effects* on transmission probability. In Section 2.2.1 we derived the Breit–Wigner formula, which expresses a Lorentzian energy dependence of the transmission probability around the resonant state:

$$T(E_z) = \frac{\Gamma^2}{(E_z - E_0)^2 + \Gamma^2} \tag{2.51}$$

where Γ is the width of the resonant level (Section 2.2.1). Jonson and Grincwajg [27] extended this formula to the sequential tunnelling regime by introducing a simple factor which expresses a random change in the phase of the wavefunction, and found that, for symmetric double barriers,

$$T(E_z) = \frac{\Gamma}{\Gamma_{\text{tot}}} \frac{\Gamma_{\text{tot}}^2}{(E_z - E_0)^2 + \Gamma_{\text{tot}}^2} \tag{2.52}$$

where Γ_{tot} is the total width of the resonant level due to both the intrinsic broadening, Γ, and the extrinsic *scattering broadening*, Γ_s: $\Gamma_{\text{tot}} = \Gamma + \Gamma_s$. It should again be noted that Γ is connected to the dwell time via eqn (2.47). In the same manner, we define the scattering-induced *phase-coherence breaking time*, τ_s, corresponding to Γ_s:

$$\tau_s = \hbar/\Gamma_s \tag{2.53}$$

Equation (2.52) shows that the ratio of Γ_s to Γ (equivalently, τ_s/τ_d) is an important factor which provides a boundary between the global coherent tunnelling and sequential tunnelling regimes. In the sequential tunnelling regime $\Gamma_s/\Gamma > 1$, the peak value of the transmission probability decreases rapidly and the transmission peak becomes much broader than that obtained from the global coherent tunnelling model. This leads to an increase in the tunnelling current in the valley regime and a dramatic decrease in the *P*/*V* current ratio. An example of the calculated results obtained from this model is shown in Fig. 2.25.

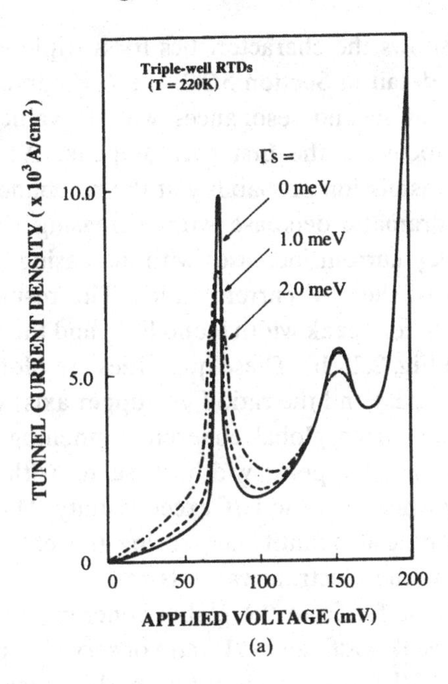

(a)

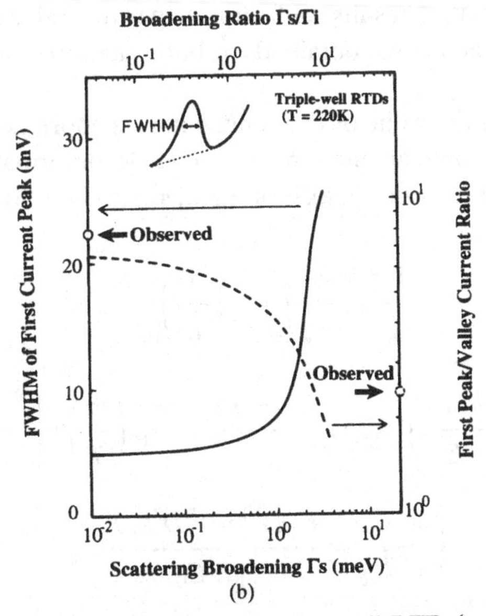

(b)

Figure 2.25 (a) *I–V* characteristics for a triple-well RTD (see Section 5.2.2 for details) around first and second resonances calculated using various values of Γ_s. (b) Γ_s-dependencies of both the first current peaks' width and *P/V* current ratios calculated at the first resonance of the triple-well RTD. Experimental data are indicated by open circles on the vertical axis.

Figure 2.25(a) shows the characteristics for a triple-well RTD which will be studied in detail in Section 5.2.2: the *I–V* curves are calculated around the first and second resonances with Γ_s values of 0, 1.0 and 2.0 meV. Let us focus on the first current peak. As a result of the decrease in the transmission probability at the resonance, the first peak current shows a dramatic decrease with increasing Γ_s. On the other hand, the first valley current increases with increasing Γ_s, resulting in a large degradation of the *P/V* current ratios. The resulting Γ_s dependencies of the first current peak width (solid line) and the first *P/V* current ratio are shown in Fig. 2.25(b). These quantities are plotted as functions both of Γ_s (bottom axis) and the ratio Γ_s/Γ (upper axis) with the Γ value of 0.3 meV obtained from global coherent tunnelling calculations. A dramatic increase in the peak width is seen, with the *P/V* ratio decreasing rapidly when the ratio Γ_s/Γ exceeds unity. This is indicative of the ratio being a crucial quantity measuring the extent to which the phase-coherence of the electron waves is maintained. If $\Gamma_s/\Gamma > 1$, the resonant tunnelling is far from the global coherent tunnelling picture. The first resonant peak width and *P/V* ratio observed experimentally are indicated in Fig. 2.25(b) by open circles on both vertical axes. Comparison of the calculated results and these experimental data shows that simultaneous agreement is obtained for both quantities with Γ_s approximately 2.5 meV.

Let us examine the value of Γ_s obtained here in more detail. By taking account of polar-optical phonon and acoustic deformation potential scattering, the momentum relaxation time, τ_m, is given by the following expressions [28]:

$$\frac{1}{\tau_m} = \left(\frac{1}{\tau_m}\right)_{LO} + \left(\frac{1}{\tau_m}\right)_{DP} \tag{2.54}$$

$$\left(\frac{1}{\tau_m}\right)_{LO} = \alpha\omega_{LO}\left(\frac{\hbar\omega_{LO}}{E}\right)^{1/2} N_q\left\{\ln\left|\frac{a+1}{a-1}\right| + \exp\left(\frac{\Theta}{T}\right)\ln\left|\frac{1+b}{1-b}\right|\right\} \tag{2.55}$$

$$\left(\frac{1}{\tau_m}\right)_{DP} = \frac{(2E)^{1/2} m^{*3/2} D^2 k_B T}{\pi\hbar^4 C_L} \tag{2.56}$$

Here α is the polar constant given by

$$\alpha = \frac{e^2 m^{*1/2}}{4\sqrt{2}\pi\kappa_0\hbar(\hbar\omega_{LO})^{1/2}}\left(\frac{1}{\kappa_{opt}} - \frac{1}{\kappa}\right) \tag{2.57}$$

Table 2.6. Phonon-induced scattering broadening derived from eqns (2.54)–(2.61)

	Momentum relaxation time τ_m(s)	Scattering broadening Γ_s(meV)
Polar-optical phonon scattering	1.92×10^{-13}	3.4
Deformation potential scattering	2.30×10^{-12}	0.3
Total	1.77×10^{-13}	3.7

where κ_0 is the permittivity of free space, κ and κ_{opt} are the static and optical dielectric constants and N_q, a, and b are defined as follows:

$$N_q = \frac{1}{\exp\left(\frac{\Theta}{T}\right) - 1} \tag{2.58}$$

$$a = \left(1 + \frac{\hbar\omega_{LO}}{E}\right)^{1/2} \tag{2.59}$$

$$b = \mathrm{Re}\left(1 - \frac{\hbar\omega_{LO}}{E}\right)^{1/2} \tag{2.60}$$

Assuming an LO-phonon energy, $\hbar\omega_{LO}$, of 36 meV, a Debye temperature, Θ, of 417 K, a polar constant, α, of 0.067, a deformation potential constant, D, of 13.5 eV, a longitudinal elastic constant, C_L, of 1.44×10^{11} Nm^{-2} and an electron energy, E, of 48 meV (the energy of the first resonant state of the triple-well RTD), the values of $(1/\tau_m)_{LO}$ and $(1/\tau_m)_{DP}$ are estimated to be 5.21×10^{12}s^{-1} and 4.34×10^{11}s^{-1} at 220 K respectively. The scattering broadening may be estimated from the momentum relaxation time through the uncertainty principle:

$$\Gamma_s \approx \hbar/\tau_m \tag{2.61}$$

Γ_s is then obtained as 3.7 meV (see Table 2.6), which is not far from the value, 2.5 meV, obtained above.

Although the above discussion is quite primitive, the results imply that the phase-coherence of electron waves is in fact partly lost during the electron dwell time in the quantum well, and the change in the transmission probability function is reasonably described by using a simple broadening model. We will study this subject more fully in the pages to come.

2.7 References

[1] M. Tsuchiya, H. Sakaki and J. Yoshino, Room temperature observation of differential negative resistance in AlAs/GaAs/AlAs resonant tunneling diodes, *Jpn. J. Appl. Phys.*, **24**, L466, 1985.

[2] M. Tsuchiya and H. Sakaki, Precise control of resonant tunneling current in AlAs/GaAs/AlAs double barrier diodes with atomically-controlled barrier widths, *Jpn. J. Appl. Phys.*, **25**, L185, 1986.

[3] M. Tsuchiya and H. Sakaki, Dependence of resonant tunneling current on well widths in AlAs/GaAs/AlAs double barrier diode structures, *Appl. Phys. Lett.*, **49**, 88, 1986.

[4] D. Landheer, G. C. Aers and Z. R. Wasilewski, Effective mass in the barriers of GaAs/AlAs resonant tunneling double barrier diodes, *Superlattices and Microstructures*, **11**, 55, 1992.

[5] A. Harwit, J. S. Harris, Jr and A. Kapitulnik, Calculated quasi-eigenstates and quasi-eigenenergies of quantum well superlattices in an applied electric field, *J. Appl. Phys.*, **60**, 3211, 1986.

[6] J. S. Blakemore, Semiconducting and other properties of gallium arsenide, *J. Appl. Phys.*, **53**, R123, 1982.

[7] B. Ricco and M. Y. Azbel, Physics of resonant tunneling. The one-dimensional double-barrier case, *Phys. Rev.*, **B29**, 1970, 1984.

[8] H. C. Liu, Tunneling time through heterojunction double barrier diodes, *Superlattices and Microstructures*, **3**, 379, 1987.

[9] D. D. Coon and H. C. Liu, Frequency limit of double barrier resonant tunneling oscillators, *Appl. Phys. Lett.*, **49**, 94, 1986.

[10] H. Ohnishi, T. Inata, S. Muto, N. Yokoyama and A. Shibatomi, Self-consistent analysis of resonant tunneling current, *Appl. Phys. Lett.*, **49**, 1248, 1986.

[11] K. F. Brennan, Self-consistent analysis of resonant tunneling in a two-barrier–one-well microstructure, *J. Appl. Phys.*, **62**, 2392, 1987.

[12] M. A. Reed, W. R. Frensley, W. M. Duncan, R. J. Matyi, A. C. Seabaugh and H. L. Tsai, Quantitative resonant tunneling spectroscopy: Current-voltage characteristics of precisely characterized resonant tunneling diodes, *Appl. Phys. Lett.*, **54**, 1256, 1989.

[13] B. Jogai, C. I. Huang and C. A. Bozada, Electron density in quantum well diodes, *J. Appl. Phys.*, **66**, 3126, 1989.

[14] E. P. Wigner, Lower limit for the energy derivative of the scattering phase shift, *Phys. Rev.*, **98**, 145, 1955.

[15] N. Harada and S. Kuroda, Lifetime of resonant state in a resonant tunneling system, *Jpn. J. Appl. Phys.* **25**, L871, 1986.

[16] M. Tsuchiya, T. Matsusue and H. Sakaki, Tunneling escape rate of electrons from quantum well in double-barrier heterostructures, *Phys. Rev. Lett.*, **59**, 2356, 1987.

[17] J. S. Wu, C. Y. Chang, C. P. Lee, K. H. Chang, D. G. Liu and D. C. Liou, Resonant tunneling of electrons from quantized levels in the accumulation layer of double-barrier heterostructures, *Appl. Phys. Lett.*, **57**, 2311, 1990.
[18] C. J. Goodings, Variable-area resonant tunnelling diodes using implanted gates, PhD thesis, Cambridge University, 1993.
[19] E. S. Alves, L. Eaves, M. Henini, O. H. Hughes, M. L. Leadbeater, F. W. Sheard, G. A. Toombs, G. Hill and M. A. Pate, Observation of intrinsic bistability in resonant tunnelling devices, *Electron. Lett.*, **24**, 1190, 1988.
[20] T. Tanoue, H. Mizuta and S. Takahashi, A triple-well resonant tunneling diode for multiple-valued logic application, *IEEE Electron Device Lett.*, **EDL-9**, 365, 1988.
[21] H. Mizuta, T. Tanoue and S. Takahashi, A new triple-well resonant tunneling diode with controllable double-negative resistance, *IEEE Trans. Electron Devices*, **ED-35**, 1951, 1988.
[22] H. Mizuta, T. Tanoue and S. Takahashi, Theoretical analysis of peak-to-valley ratio degradation caused by scattering processes in multi-barrier resonant tunneling diodes, *Proceedings of IEEE/Cornell Conference on Advanced Concepts in High Speed Semiconductor Devices and Circuits*, p. 274, 1989.
[23] T. Nakagawa, H. Inamoto, T. Kojima and K. Ohta, Observation of resonant tunneling in AlGaAs/GaAs triple barrier diodes, *Appl. Phys. Lett.*, **49**, 73, 1986.
[24] S. Luryi, Frequency limit of double-barrier resonant-tunneling oscillators, *Appl. Phys. Lett.*, **47**, 490, 1985.
[25] S. Luryi, Coherent versus incoherent resonant tunneling and implications for fast devices, *Superlattices and Microstructures*, **5**, 375, 1989.
[26] T. Weil and B. Vinter, Equivalence between resonant tunnelling and sequential tunnelling in double-barrier diodes, *Appl. Phys. Lett.*, **50**, 1281, 1987.
[27] M. Johnson and A. Grincwajg, Effect of inelastic scattering on resonant and sequential tunneling in double barrier heterostructures, *Appl. Phys. Lett.*, **51**, 1729, 1987.
[28] See, for example, K. Seeger, *Semiconductor Physics An Introduction* (Springer Series in S. S. Science, Berlin, 1985).

3
Scattering-assisted resonant tunnelling

In Section 2.6 we briefly studied the effects of electron scattering on resonant tunnelling which are inevitable in a real system operating at room temperature. The phenomenological Breit–Wigner formula was introduced to describe the incoherent aspect of the electron tunnelling which in general results in a broadening of the transmission peak and thus degraded current *P*/*V* ratios in RTDs. In this chapter we look in more detail at various scattering processes, both elastic and inelastic, which have been of great interest not only from a quantum transport physics point of view but also because of the possibility of controlling and even engineering these interactions in semiconductor microstructures. The inelastic longitudinal–optical (LO) phonon scattering, introduced in the preceding chapter, is the most influential process, with Γ–X-intervalley scattering and impurity scattering also affecting the resonant tunnelling electrons. Section 3.1 describes the dominant electron–LO-phonon interactions. Both theoretical and experimental studies of a postresonant current peak are presented, which provide much information about the electron–phonon interactions in the quantum well. Section 3.2 then discusses the effects of the upper X-valley which become more significant in $Al_xGa_{1-x}As/GaAs$ systems with an Al mole fraction, x, higher than 0.45 since the energy of the X-valley then becomes lower than that of the Γ-valley. Finally, in Section 3.3, we study elastic impurity scattering, which may be caused by residual background impurities or those diffused from the heavily doped contact regions.

3.1 LO-phonon-emission-assisted resonant tunnelling

3.1.1 Phonon-assisted tunnelling and postresonance peak

Interactions between electrons and LO-phonons are in general the most important energy dissipation processes in polar semiconductor systems. The electrons in the double-barrier structure may emit or absorb LO-phonons during the resonant tunnelling process. As this is an inelastic scattering process the electrons involved do not conserve the components of momentum either parallel or perpendicular to the barriers. Resonant tunnelling accompanied by a single LO-phonon emission is called *LO-phonon-assisted resonant tunnelling* (see Fig. 3.1(a)). The probability of this process is relatively small compared with that of the main resonant tunnelling process, but the result can be seen directly in the *I–V* characteristics at low temperatures as a shallow satellite current peak in the postresonance valley current (see Fig. 3.1(b)). Higher-order processes mediated by multiple-phonon emission are normally very rare, and their contributions to the *I–V* characteristics are virtually zero.

Longitudinal-optical phonon-assisted resonant tunnelling was first observed experimentally by Goldman *et al.* [2] using an $Al_{0.4}Ga_{0.6}As$(8.5 nm)/GaAs(5.6 nm)/$Al_{0.4}Ga_{0.6}$(8.5 nm) double-barrier RTD. They found a small current peak in the valley region at 4.2 K with a magnitude of about 0.04 of the main peak. The measured voltage separation between the main and satellite peaks is about 90 mV and thus a quasi-bound state in the well is thought to be located about 45 meV below the quasi-Fermi level in the emitter region. This energy is fairly close to the energy of LO-phonons in an AlAs-like interface mode (explained later in this section) and so this peak was interpreted as being due to a single LO-phonon-emission-assisted resonant tunnelling process.

The LO-phonon-assisted resonant tunnelling is particularly pronounced in RTDs in which the emitter states are quantised (as described in Section 2.4) because the main current peak becomes sharper and the intrinsic valley current is sufficiently small to show the low-satellite current peak. Chevoir and Vinter [3] have shown theoretically that the existence of such a peak, well separated from the main resonance, is a consequence of a localised emitter state. Both Materials 2 and 3 introduced in Section 2.4 show distinct, broad satellite current peaks in their *I–V* characteristics after the main resonance. The *I–V* characteristics for Material 3 are shown in Fig. 3.1(b) for an extended bias sweep. The effects of LO-phonons are found to be far stronger in Material 3

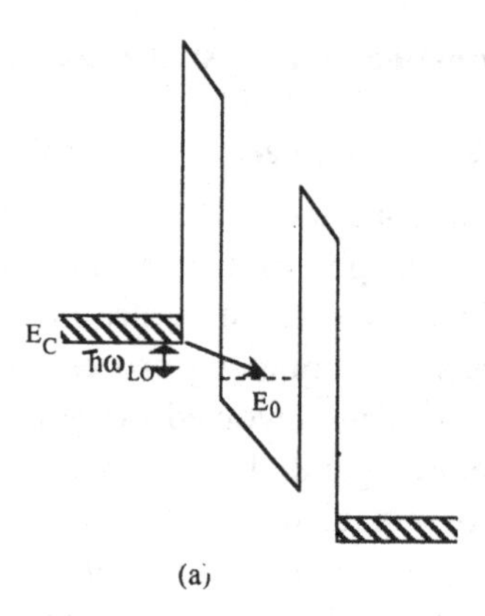

(a)

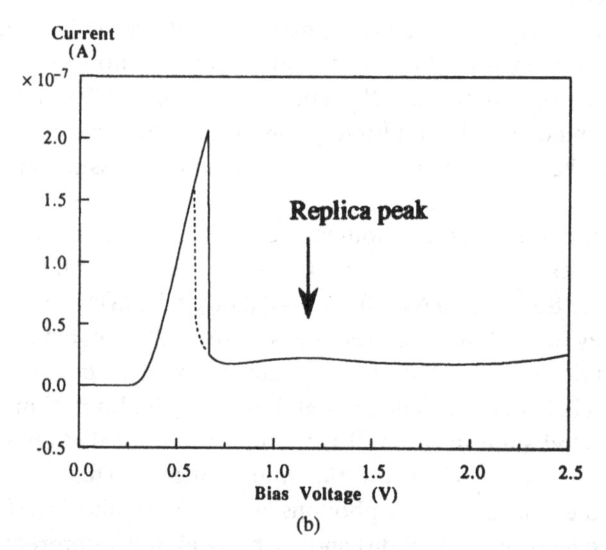

(b)

Figure 3.1 (a) The LO-phonon emission-assisted resonant tunnelling process. (b) *I–V* characteristics of an AlAs/GaAs/AlAs double-barrier RTD (Material 3; see Section 2.4 for details) measured at 4.2 K. A shallow broad peak after the main resonance is caused by LO-phonon emission-assisted tunnelling. After Goodings [1], with permission.

than in Material 2, probably due to the longer dwell time in Material 3 giving a greater electron–phonon interaction probability in the double-barrier structure. Thus Material 3 will be used in Section 3.1.3 for the purpose of further study of LO-phonon-assisted tunnelling in the presence of a magnetic field.

3.1.2 Theoretical investigations of phonon-assisted tunnelling

The theory of inelastic phonon-assisted tunnelling is much more difficult than the global coherent tunnelling theory of Chapter 2, and quite a few studies have recently been reported on this subject. In general, scattering processes in resonant tunnelling structures affect both the carrier transmission probability and their distribution in energy space. Longitudinal–optical phonon-assisted resonant tunnelling may be described theoretically using a modified transmission probability function, $T(\mathbf{k}_{/\!/},k_{\perp})$, which now depends on the momentum components perpendicular and parallel to the barriers. Here we introduce two recent theoretical studies which investigate the effects of LO-phonon scattering on the transmission probability function. The influences on the distribution function are studied in Chapter 4.

A direct way of obtaining the modified transmission probability is to sum all the transition probabilities via LO-phonons between the unperturbed coherent states. Chevoir and Vinter [4] have performed 3D numerical calculations of the following scattering rates due to the Fröhlich Hamiltonian, $H_{\text{e-ph}}$, [5] (see also eqn (3.3)) for electron–LO-phonon interactions:

$$S(\mathbf{K},\mathbf{K}') = \frac{2\pi}{h}|\langle\mathbf{K}'|H_{\text{e-ph}}|\mathbf{K}\rangle|^2(1 - f(\mathbf{K}'))\delta(E_{\mathbf{K}'} - E_{\mathbf{K}} - \hbar\omega_{\text{LO}}) \qquad (3.1)$$

by using a complete set of coherent scattering states denoted by $\mathbf{K} = (\mathbf{k}_{/\!/},k_{\perp})$, calculated for the whole double-barrier structure. The transmission probability of phonon-assisted tunnelling is then calculated as follows:

$$T(\mathbf{K}) \approx W(\mathbf{K}) = \frac{Lm^*}{\hbar k_{\perp}}\sum_{\mathbf{K}'} S(\mathbf{K},\mathbf{K}') \qquad (3.2)$$

where L is the total length of the device. The transmission probability function (3.2) includes contributions from all the *real* scattering processes, and it has been shown by Chevoir and Vinter that the phonon replica peak is well reproduced in the I–V characteristics calculated from eqns (2.40) and (3.2). This simple and practical method is suitable for large-scale numerical calculations, although it excludes the important effects of *virtual* scattering processes which result in the breaking of phase-coherence [4].

Another theoretical method is based on the transfer Hamiltonian formula, studied in Section 2.2.3, which goes into more of the detailed physics of LO-phonon-assisted resonant tunnelling. Several theoretical studies have been reported by Wingreen *et al.* [6], Jonson *et al.* [7], Zheng *et al.* [8], Turley *et al.* [9], [10] and Mori *et al.* [11]. As mentioned earlier, the transfer Hamiltonian formula is preferably used for this as the many-body effects in the quantum well can be incorporated via the self-energy of electrons in the well. In this method electron–phonon interactions are introduced as effective electron–electron interactions due to the virtual exchange of LO-phonons.

The Fröhlich Hamiltonian for the electron–phonon interactions in layered structures, $H_{\text{e-ph}}$, is expressed as follows [5]:

$$H_{\text{e-ph}} = \frac{1}{\sqrt{V}} \sum_{j,\mathbf{q}_{\parallel},q_{\perp},\mathbf{k}_{\parallel}} M_j(\mathbf{q}_{\parallel},q_{\perp}) a^{+}_{\mathbf{k}_{\parallel}+\mathbf{q}_{\parallel},k^0_{\perp}} a_{\mathbf{k}_{\parallel},k^0_{\perp}} (b_{j,\mathbf{q}_{\parallel},q_{\perp}} + b^{+}_{j,-\mathbf{q}_{\parallel},-q_{\perp}}) \quad (3.3)$$

where the matrix element $M_j(\mathbf{q}_{\parallel},q_{\perp})$ represents the interaction between the electrons and the confined- and interface-phonons given by the following expressions:

$$M_j(\mathbf{q}_{\parallel},q_{\perp}) = \mathrm{i}\left\{\frac{2\pi\hbar\alpha_j}{q_{\parallel}^2 + q_{\perp}^2}\left(\frac{2(\hbar\omega_j)^3}{m^*}\right)^{\frac{1}{2}} F_j\right\}^{\frac{1}{2}} \quad (3.4)$$

The index j in eqns (3.3) and (3.4) indicates different phonon modes. For symmetric AlAs/GaAs/AlAs double-barrier structures these modes are the confined LO mode (denoted by c), the GaAs-like symmetric interface modes (denoted by $s\pm$ (GaAs)) and the AlAs-like symmetric interface modes (denoted by $s\pm$ (AlAs)). Figure 3.2 shows the dispersion curves for different optical-phonon modes calculated by Mori *et al.* [11] using the dielectric continuum model for an AlAs/GaAs/AlAs double-barrier structure with a well width of 5 nm and barrier thicknesses of (a) 10 nm and (b) 2.5 nm.

The two branches shown by solid and broken lines are the GaAs-like interface modes, and two others shown by single- and double-dotted chain lines are the AlAs-like interface modes. The confined LO-phonon mode is indicated by dotted lines. In eqn (3.4) the mode-dependent Fröhlich coupling constants, α_j ($j = c$ and $s\pm$), are defined by

$$\alpha_c = \frac{e^2}{4\pi\epsilon_0\hbar}\left(\frac{m^*}{2\hbar\omega_c}\right)^{\frac{1}{2}} \beta_w(\omega_c) \quad (3.5)$$

for confined LO-phonon modes, and

$$\alpha_s^{\pm} = \frac{e^2}{4\pi\epsilon_0\hbar}\left(\frac{m^*}{2\hbar\omega_{s\pm}}\right)^{\frac{1}{2}} \frac{D^{\pm} + 1}{D^{\pm}\beta_w^{-1}(\omega_{s\pm}) + \beta_b^{-1}(\omega_{s\pm})} \quad (3.6)$$

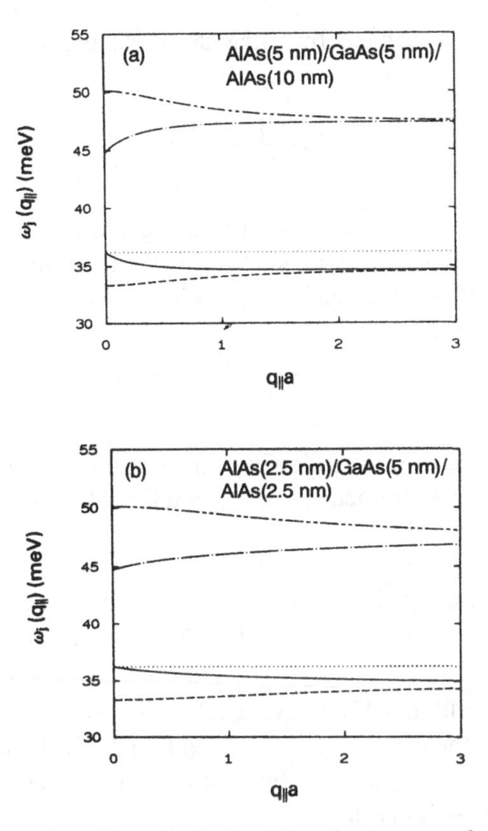

Figure 3.2 Frequencies of symmetric interface phonons as a function of $q_{\parallel}a$ for AlAs/GaAs/AlAs symmetric double-barrier structures (a) $a = 5$ nm, $b = 10$ nm and (b) $a = 5$ nm, $b = 2.5$ nm. The solid curves are the S_0^+ (GaAs) mode, the broken curves the S_0^- (GaAs) mode, the dotted broken curves the S_1^+ (AlAs) mode and the double-dotted broken curves the S_1^- (AlAs) mode. The dotted curves are the confined LO mode. After Mori *et al.* [11], with permission.

for symmetric interface modes, where the indices w and b indicate the well (GaAs) and barrier (AlAs) materials, and

$$D^{\pm} = \frac{1+A}{2B}\left\{1 \pm \sqrt{\left(1 - \left(\frac{2B}{1+A}A\right)^2\right)}\right\} \tag{3.7}$$

$$A = \tanh(q_{\parallel}a/2) \tag{3.8}$$

$$B = \tanh(q_{\parallel}b) \tag{3.9}$$

$\beta_n(\omega)$ ($n = w,b$) is defined by using the dielectric functions $\epsilon_n(\omega)$ as follows:

$$\beta_n(\omega)\frac{1}{\omega^2}\left|\frac{\partial\epsilon_n(\omega)}{\partial\omega^2}\right|^{-1} \tag{3.10}$$

When the Hamiltonian, $H_{\text{e-ph}}$, is introduced into the total Hamiltonian in eqn (2.30), the self-energy of electrons in the quantum well, Σ_{w}, in eqn (2.37) consists of contributions from both the tunnelling and electron–phonon interactions:

$$\Sigma_{\text{w}} = \Sigma_{\text{w}}^{\text{t}} + \Sigma_{\text{w}}^{\text{e-ph}} \tag{3.11}$$

Mori *et al.* have shown the following expression for the self-energy due to electron–phonon interactions, $\Sigma_{\text{w}}^{\text{e-ph}}$, by taking the lowest term of $H_{\text{e-ph}}$ into consideration [11]:

$$\Sigma_{\text{w}}^{\text{e-ph}}(\mathbf{k}_{/\!/}^{\text{w}},k_{\perp}^{0},E) = \sum_{j}\sum_{\mathbf{q}_{/\!/}}\frac{\hbar\alpha_j(\hbar\omega_j)^{3/2}F_j}{\sqrt{m^*}}\,\frac{1}{E-\hbar\omega_j-\epsilon_{\mathbf{k}_{/\!/}-\mathbf{q}_{/\!/}}-\epsilon_{k_{\perp}^{0}}+\text{i}\Gamma} \tag{3.12}$$

where the tunnelling self-energy, $\Sigma_{\text{w}}^{\text{t}}$, has simply been replaced by a constant $-\text{i}\Gamma$ in the same manner as in Section 2.2.3. The form factor, F_j, in eqns (3.4) and (3.12) is given by the following expressions depending on the phonon modes [11]:

$$F_{\text{c}}(q_{/\!/}) = \int_{-\infty}^{\infty}\text{d}z\int_{-\infty}^{\infty}\text{d}z'|\varphi_{k_{\perp}^{0}}(z)|^2|\varphi_{k_{\perp}^{0}}(z')|^2$$

$$\times\left\{\text{e}^{-q_{/\!/}|z-z'|}-\frac{1}{1+\tanh(q_{/\!/}a/2)}\,\frac{\cosh(q_{/\!/}z)\cosh(q_{/\!/}z')}{\cosh^2(q_{/\!/}a/2)}\right\} \tag{3.13}$$

for confined LO-phonon modes, and

$$F_{\text{s}}^{\pm}(q_{/\!/}) = \frac{1}{\Lambda^{\pm}(D^{\pm}+1)}\left\{\int_{-\infty}^{\infty}\text{d}z\frac{\cosh(q_{/\!/}z)}{\cosh(q_{/\!/}a/2)}|\varphi_{k_{\perp}^{0}}(z)|^2\right\}^2 \tag{3.14}$$

for symmetric interface modes, where $\varphi_{k_{\perp}^{0}}(z)$ is the electron wavefunction at the resonant state, and

$$\Lambda^{\pm} = \frac{1}{2}(1-B)B\left(\frac{D^{\pm}+1}{D^{\pm}-B}\right)^2+\frac{1}{2}(1+B)B\left(\frac{D^{\pm}-1}{D^{\pm}-B}\right)^2 \tag{3.15}$$

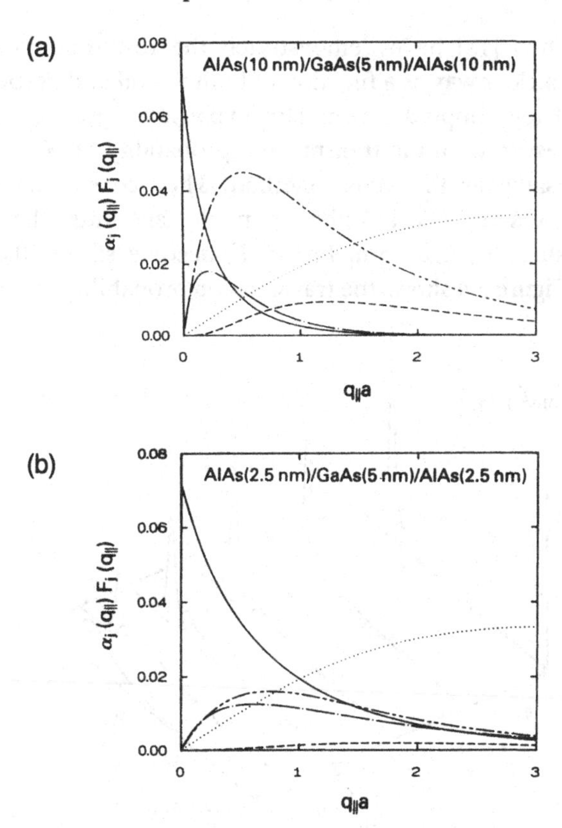

Figure 3.3 Products of coupling constants and form factors in AlAs/GaAs/AlAs symmetric double-barrier structures as a function of $q_{\parallel}a$. (a) $a = 5$ nm, $b = 10$ nm and (b) $a = 5$ nm, $b = 2.5$ nm. The solid curves are the S_0^+ mode, the broken curves the S_0^- (GaAs) mode, the dotted broken curves the S_1^+ (AlAs) mode and the double-dotted broken curves the S_1^- (AlAs) mode. The dotted curves are the confined LO mode. After Mori *et al.* [11], with permission.

The resulting coefficient, $\alpha_j F_j$, in eqn (3.12) is shown in Fig. 3.3, calculated by Mori *et al.* [11] as a function of $q_{\parallel}a$ for the two tunnelling barrier structures. In both cases the GaAs-like symmetric interface mode is dominant in the long wavelength limits. In the case of thicker barriers (Fig. 3.3(a)) the AlAs-like symmetric interface mode contributes most in the region $0.5 < q_{\parallel}a < 1.5$, with the GaAs confined mode becoming dominant for $q_{\parallel}a > 1.5$. The AlAs-like modes play less of a role in the case of thinner barriers (Fig. 3.3(b)) and the GaAs-like modes are dominant throughout the whole region. This indicates that the phonon modes in the thinner structure are similar to those in bulk material.

Zou and Chao [12] have demonstrated the resulting transmission probability in a clear way as a function of both lateral and perpendicular momentum. They adopted a cumulant expansion approach to calculating the lowest term in the transmission probability of $H_{e\text{-}ph}$, which is basically the same as the above method. They considered only the dispersionless GaAs bulk LO-phonon mode, and thus the Fröhlich coupling constant, α, and form factor, F, become simply 0.07 and 1 respectively. Figure 3.4 shows the transmission probability calculated for

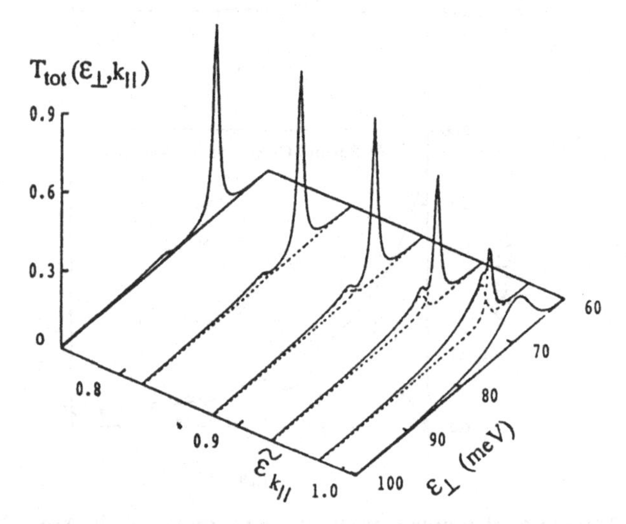

Figure 3.4 The total transmission probability T_{tot} as a function of the perpendicular energy of an emitted electron for various values of parallel energy at zero bias and at zero temperature. Parallel energies $\epsilon_{k_{/\!/}}$ are normalised by using the LO-phonon energy $\hbar\omega_{LO}$ of 36 meV. After Zou and Chao [12], with permission.

an $Al_{0.3}Ga_{0.7}As/GaAs/Al_{0.3}Ga_{0.7}As$ (4 nm/5 nm/4 nm) double-barrier structure at zero temperature. The lateral and perpendicular energies are normalised using a GaAs LO-phonon energy of 36 meV. For small lateral electron energies, a shallow replica peak due to LO-phonon-emission-assisted tunnelling is found, this being well separated from the sharp main resonance peak. It can be clearly seen that the replica peak moves towards the main one with increasing lateral energy and eventually merges into it for higher lateral energies, $\epsilon_{k_{/\!/}} \approx \hbar\omega_{LO}$. It can also be seen that the effects of the phase-coherence breaking process are enhanced in this region, resulting in the dramatic collisional broadening of the transmission peak which we have studied in Section 2.6 using the phenomenological model. These results indicate that the contribution of the LO-phonon-assisted tunnelling process is likely to be washed out in tunnelling structures with high Fermi energy in the emitter region.

Mori *et al.* [11] have adopted eqns (2.37), (2.40) and (3.12), including the polaron energy shift [13], to calculate the *I–V* characteristics for two types of AlAs/GaAs/AlAs symmetric double-barrier structure. Figure 3.5 shows the characteristic for (a) a structure with thicker AlAs barriers of 10 nm and (b) one with thinner barriers of 2.5 nm. The solid line is the current calculated by taking account of the electron–interface–phonon interactions. For comparison, also shown are the results obtained by assuming the electron–bulk–phonon interactions (the broken line) and no electron–phonon interaction (the dotted line).

In the case of the thicker barriers (Fig. 3.5(a)) the solid line shows two satellite peaks associated with the GaAs-like (at a lower peak voltage) and AlAs-like (at a higher peak voltage) interface–phonon-emission-assisted resonant tunnelling processes. The confined phonon mode contributes to the results much less than these since electrons interact mainly with phonons with $q < k_F$. In the case of the thinner barriers (Fig. 3.5(b)), the second satellite peak due to the AlAs-like interface phonon becomes much smaller and the overall curve is similar to the results calculated with the bulk phonon model (the broken line). The change in the detail of the satellite peaks has been, at least quantitatively, confirmed by Leadbeater *et al.* [14], who observed a flat-top satellite structure for a device with relatively thick asymmetric double barriers. In the presence of a magnetic field this peak is clearly resolved into two peaks which are attributable to the interface modes of AlGaAs barriers.

3.1.3 Magnetotunnelling measurements in the valley-current regime

Let us move on to the experimental study of the LO-phonon-assisted replica peak using magnetic fields. The use of magnetic fields is an important tool in semiconductor research, and the field of RTDs is by no means an exception to this, with various effects being seen for magnetic fields applied both perpendicular and parallel to the barriers. In particular, for a field perpendicular to the barriers, effects have been seen on the elastic tunnelling on-resonance and the inelastic tunnelling off-resonance, which have been used to provide information about LO-phonon scattering, as well as impurity scattering (see Section 3.3), space

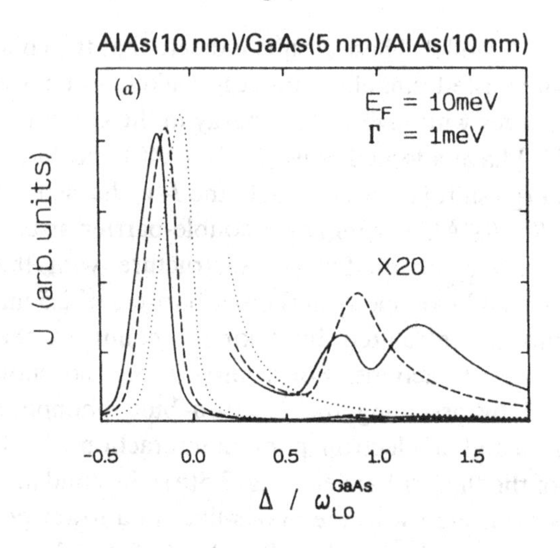

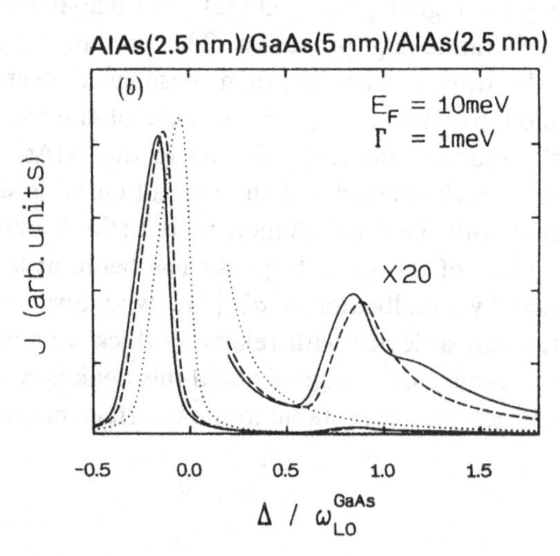

Figure 3.5 *I–V* characteristics for GaAs/AlAs symmetric double-barrier structures. (a) $a = 5$ nm, $b = 10$ nm and (b) $a = 5$ nm, $b = 2.5$ nm. Δ is the difference between the bottom of the emitter and the resonant level. The solid and broken curves include the electron–interface–phonon interactions and the electron–bulk–photon interactions. The dotted curves are without any phonon effects. After Mori *et al.* [11], with permission.

charge build-up (see Section 4.3.2) and related intrinsic bistability (see Section 4.4).

We are now concerned with the tunnelling processes beyond the main resonance, and for these the component of momentum perpendicular to the barrier is no longer conserved. A magnetic field, B, applied perpendicular to the tunnelling barriers quantises the lateral motion of electrons in both the emitter and well regions. The transverse kinetic energy $\hbar^2 k_{\|}^2/2m^*$ is replaced by the Landau subband energy, and the total electron energy, E is expressed in both regions as follows:

$$E = E_0 + (N + 1/2)\hbar\omega_c \tag{3.16}$$

where E_0 is the lowest 2D state energy, N is the Landau-level quantum number, and ω_c is the cyclotron frequency ($=e\hbar B/m^*$). Energy conservation at the resonance, with or without phonon emission, is then given by the following expression:

$$E_0(\text{emitter}) + (N + 1/2)\hbar\omega_c = E_0(\text{quantum well}) + (N' + 1/2)\hbar\omega_c + m\hbar\omega_{LO} \tag{3.17}$$

where m is zero or unity since the multiple-phonon-assisted tunnelling process is exceedingly rare. If $m = 0$, tunnelling is elastic without phonon emission, and if $m = 1$, tunnelling is inelastic. Any difference between N and N' represents the change in Landau level during tunnelling. The result of eqn (3.17) is a series of periodic structures in the valley current at finite magnetic fields. A plot of the peak voltages of these structures as a function of the field should give a very characteristic fan diagram [14] in which the angle between the spokes of the fan depends on the spacing of the Landau levels, and the intercept depends on the difference in the energy levels and the LO-phonon energy.

Characteristics for the valley current of Material 3 are shown in Fig. 3.6 [1], in which successive field traces have been displaced by 1 nA for clarity. At zero field, the device shows a broad peak after the main resonance peak. It has been found that the application of a magnetic field enhances and splits the zero-field feature, and this provides a method of analysing the types of interaction that occur.

Peak data was extracted from Fig. 3.6 by two methods. The first involved taking the derivative of the data to obtain the conductance and then looking for zero-crossings or points of inflection. This was useful for both the relatively large peaks and those at low fields. Alternatively, the gross zero-field data was subtracted from the higher field results, enabling small features otherwise masked by the large zero-field peak to be extracted. Both sets of data are combined in Fig. 3.7 [15].

Various definite trends can be seen on the fan diagram, as indicated

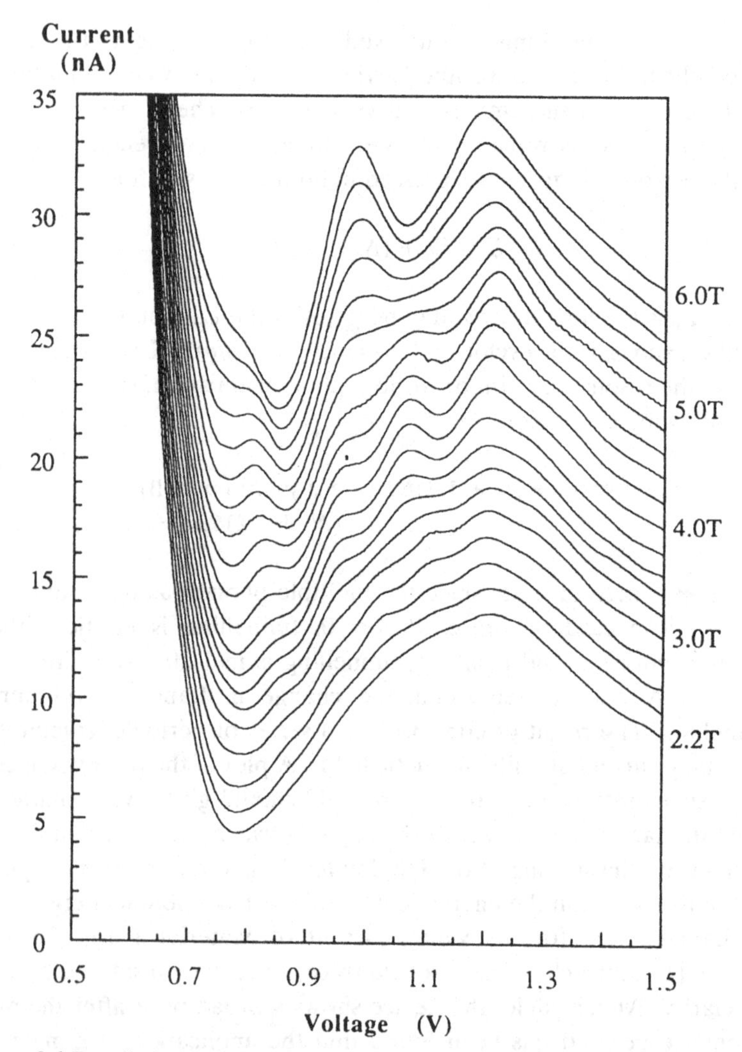

Figure 3.6 Detailed view of the valley current for a 14 μm diameter device of Material 3 at 4.2 K for various magnetic fields perpendicular to the barriers. The scale corresponds to the 2.2 T data with other curves offset by 1 nA intervals for clarity. The interval between each of the curves is 0.2 T. After Goodings [1], with permission.

by the lines in Fig. 3.7, which have been added in accordance with the energy conservation equation, eqn (3.17). Of vital importance to the understanding of such a diagram is the correct assignment of the energy scale corresponding to the applied bias voltage, and we will assume here that a linear relationship holds. According to eqn (3.17), at zero field the

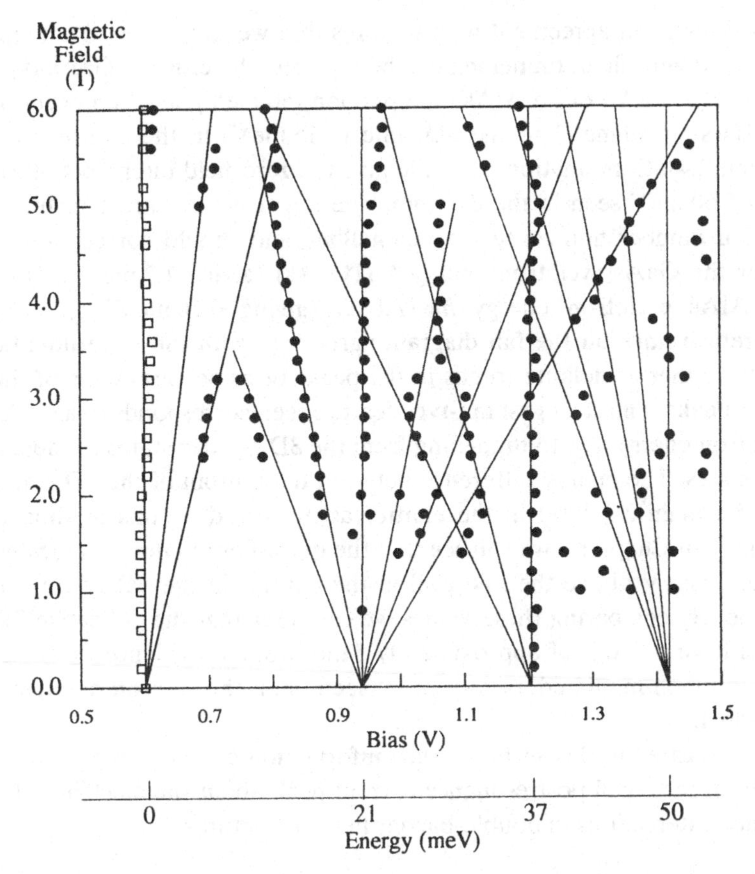

Figure 3.7 Fan diagram for phonon-assisted tunnelling in Material 3. Data points show the position of peaks in the *I–V* characteristic plotted as a function of magnetic fields applied perpendicular to the barriers. Lines are plotted according to the required energy conservation equation, eqn (3.17). The position of the main resonance peak is given by the open squares. After Goodings *et al.* [15], with permission.

only feature arises from an LO-phonon-assisted event. Therefore, the zero-field feature (at 1.2 V) should correspond to an energy of either 37 meV (GaAs-type phonon) or 50 meV (AlAs-type phonon), with the zero in the energy scale occurring at the peak resonance voltage. By considering the angles of the fans seen in the diagram we can show that the assignment of 37 meV to this feature is consistent with the other structures. The resulting energy scale is plotted beneath the bias axis in Fig. 3.7.

The structure seen in Fig. 3.7 is at first sight quite confusing, but all the lines show good agreement with features that we can expect from eqn (3.17). At zero field, tunnelling can be mediated by either a GaAs-type phonon (37 meV) or an AlAs-type phonon (50 meV), and can occur at the 2D state (0 meV) or the 3D state (−15 meV) in the emitter (see Section 2.4). Combinations of these give the zero-field intercepts of 21, 37 and 50 meV seen in the diagram. The fan structures are a result of Landau number non-conserving tunnelling, and should correspond to either the GaAs cyclotron energy $\hbar\omega_c$(GaAs) (giving 1.7 meV T^{-1}) or the AlAs cyclotron energy $\hbar\omega_c$(AlAs) (giving 0.76 meV T^{-1}). The theoretical lines on the fan diagram agree well with the experimental results, either as definite trends in the peaks or as perturbations of the larger peaks. The strongest fan-type feature seen corresponds to a GaAs cyclotron energy and to tunnelling from the 3D emitter states to the 2D well states. The energy difference between the bottom of the 3D states and the main 2D level in the emitter as measured by this method is 15 meV. In Chapter 4 we will see that the quasi-Fermi energy is located about 20 meV above the 2D level in the emitter in the valley current regime. By comparing these values we can infer that the collective 3D states have a width of approximately 5 meV. Some evidence of AlAs-type phonon-mediated processes is seen, but this is comparatively indistinct.

As indicated in this section, much information can indeed be extracted from the small postresonance current peak about the electron–LO-phonon interactions in double-barrier heterostructures.

3.2 Resonant tunnelling through X-point states

Up to now we have looked at electrons in the Γ-valley without paying attention to the contributions from those in the upper X- and L-valleys. This is reasonable as long as these upper valleys are at much higher energies than the Γ-valley. This is the case for $Al_xGa_{1-x}As$ barriers with $x < 0.45$, where the Γ-valley forms the conduction-band minimum. With increasing x the energy of the bottom of the Γ-valley increases, while that of the X-valley decreases and the bottom of the X-valley forms the conduction-band minimum for $x > 0.45$: the $Al_xGa_{1-x}As$ is then an indirect-gap semiconductor. Thus if we draw the X-band profile of $Al_xGa_{1-x}As$/GaAs/$Al_xGa_{1-x}As$ heterostructures over the Γ-band profile it is found that the X-valley of GaAs acts as a barrier and the X-valley of $Al_xGa_{1-x}As$ forms a quantum well. Figure 3.8 shows the band diagrams for $Al_xGa_{1-x}As$/GaAs/$Al_xGa_{1-x}As$ heterostructures (a) with $x = 0.4$ and

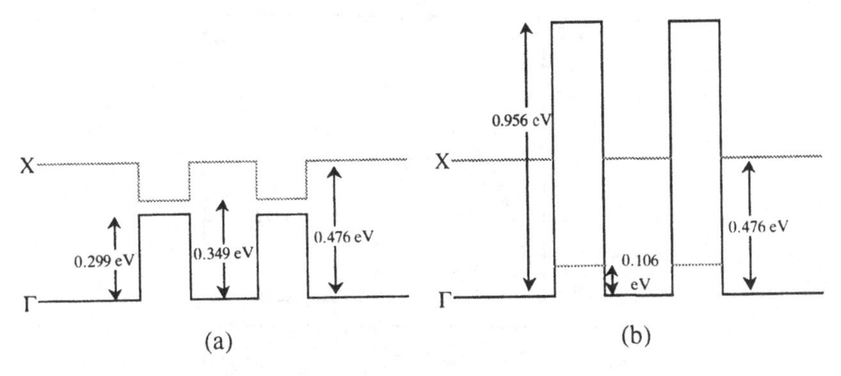

Figure 3.8 Γ- and X-band profiles for (a) $Al_{0.4}Ga_{0.6}As/GaAs/Al_{0.4}Ga_{0.6}As$ and (b) AlAs/GaAs/AlAs double-barrier structures.

(b) with $x = 1.0$. In the AlAs/GaAs/AlAs structure (Fig. 3.8(b)) the Γ(GaAs)/X(AlAs) discontinuity is only 0.106 eV, which is much smaller than the Γ(GaAs)/Γ(AlAs) separation of 0.956 eV.

3.2.1 *Resonant* Γ–X *intervalley tunnelling*

The idea of resonant tunnelling through the X-valley was first reported by Mendez *et al.* [16] in an $Al_{0.4}Ga_{0.6}As/GaAs/Al_{0.4}Ga_{0.6}As$ double-barrier structure. As shown in Fig. 3.8(a) the X-band is located just above the Γ-band in this structure. Figure 3.9 shows the *I–V* characteristics for RTDs with 10-nm-thick $Al_{0.4}Ga_{0.6}As$ barriers and a GaAs well of (a) 4 nm and (b) 10 nm thickness. Mendez *et al.* observed weak current shoulders at higher applied voltages (≈1.0 V) in both structures, in addition to the clear negative differential conductance that is attributable to conventional Γ-band resonant tunnelling. These extra structures could not be explained by tunnelling through virtual states above the Γ-band and were attributed to tunnelling through the upper X-valley.

The Γ–X-intervalley transition in such structures is an *elastic* intervalley process resulting from *Γ–X-band-mixing* at the AlGaAs/GaAs heterointerface rather than a conventional phonon-mediated intervalley scattering seen in bulk materials at high fields. Hereafter, we call this process *resonant intervalley tunnelling*. This should conserve momentum components parallel to the heterointerface: electrons are transferred from the GaAs Γ-valley to the (100) ellipsoid of the AlGaAs X-valley. Thus the associated electron effective mass in the AlGaAs valley is the heavy longitudinal mass, m_l^*, and this tunnelling should be distinguished

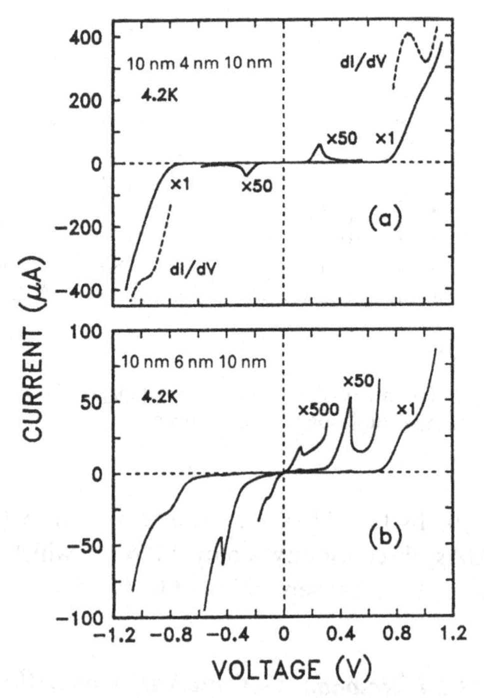

Figure 3.9 Experimental I–V characteristics of $Ga_{1-x}Al_xAs/GaAs/Ga_{1-x}Al_xAs$ heterostructures with identical barriers of 10 nm and a quantum well of either (a) 4 nm or (b) 6 nm thickness. In (a), the conductance dI/dV is also shown at high bias, emphasising the weak resonant tunnelling structure corresponding to energies above the Γ-point barrier. After Mendez *et al.* [16], with permission.

from other Γ–X intervalley scattering processes which do not conserve lateral momentum.

A proper theoretical description of Γ–X resonant intervalley tunnelling obviously requires microscopic band-structure calculations beyond the *envelope function* calculations that we have used so far. Some theoretical work has been reported on this issue, using pseudo-potential [17], [18] and tight-binding [19], [20] calculations. Calculations of the I–V characteristics, including the contributions of the Γ- and X-valleys, were first undertaken by Cade *et al.* [20] for an asymmetric AlAs/GaAs superlattice based on a nearest-neighbour tight-binding scheme. Also, the extended transfer matrix calculation, in which Γ–X-mixing is simply modelled by using an empirical parameter, has been reported by Liu [21] for single AlAs barrier structures. Mendez *et al.* [16] assumed a finite amplitude for the Γ–X-intervalley transition of electrons at the

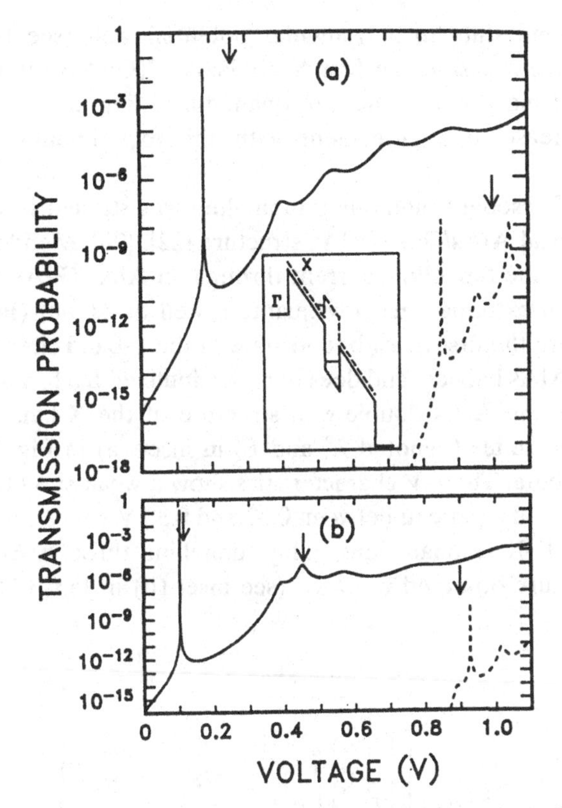

Figure 3.10 Calculated transmission probability vs applied voltage for the same two samples as those given in Fig. 3.9. Continuous and discontinuous lines correspond to intravalley tunnelling (through a $\Gamma \rightarrow \Gamma \rightarrow \Gamma \rightarrow \Gamma \rightarrow \Gamma$ path) and intervalley tunnelling (through a $\Gamma \rightarrow X \rightarrow X \rightarrow \Gamma \rightarrow \Gamma$ path) respectively. The two different potential profiles, for the 2 nm configuration, are sketched in the inset of (a). The arrows indicate the positions of the experimental conductance minima. After Mendez *et al.* [16], with permission.

heterointerfaces in the structures and calculated the voltage dependence of the Γ–X resonant intervalley tunnelling probability (see Fig. 3.10). In this diagram the solid lines represent the transmission probability of the conventional Γ-band tunnelling for electrons at the Γ-band in the emitter region and the arrows indicate the experimentally observed peak voltages. The broken lines show the transmission probability through one of the intervalley tunnelling paths, Γ(GaAs)–X(AlGaAs)–X(GaAs)–Γ(AlGaAs)–Γ(GaAs). For both structures no resonance can be seen in the Γ-band tunnelling at around 0.9 V except for weak oscillations due to the virtual states. On the other hand, it was found that a quasi-bound state in a triangular potential well (see the inset of Fig. 3.10(a)) formed at the first X(AlGaAs)/X(GaAs) interface under an external bias opens a new resonant tunnelling channel at around 0.9 V, in reasonable agreement with the experimentally observed peaks.

The Γ–X resonant intervalley tunnelling was studied in more detail using a special AlAs/GaAs/AlAs structure [22], [23]. Mendez *et al.* [22] investigated the tunnelling current through an AlAs/GaAs/AlAs structure with an extremely narrow quantum well of 2.0 nm (Fig. 3.11). In this structure the first quasi-bound state in the Γ-band rises to near the top of the AlAs barriers and does not contribute with a low applied bias. In contrast, the AlAs double-well structure of the X-band forms two quasi-bound states (denoted E_1^x and E_2^x in inset (a) in Fig. 3.11) which may contribute. The *I–V* characteristics show a weak shoulder at a bias of ~0.17 V and a plateau between 0.32 and 0.38 V which are thought to arise from Γ–X resonant intervalley tunnelling through E_1^x and E_2^x. A weak structure observed in d*I*/d*V* (see inset (b) in Fig. 3.11) at a very

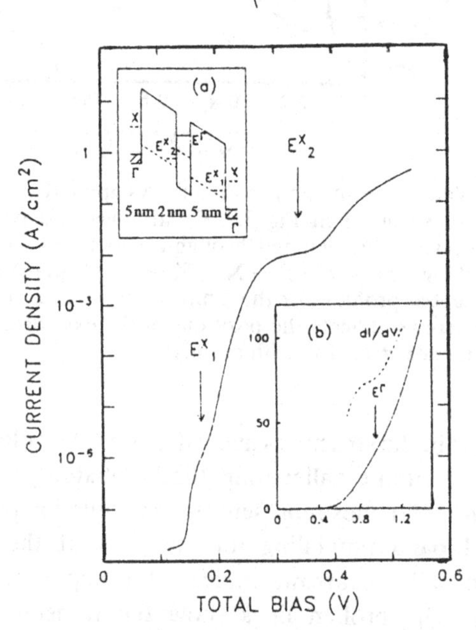

Figure 3.11 Tunnelling current (logarithmic scale) for an AlAs(5 nm)/GaAs(2 nm)/AlAs(5 nm) device as a function of voltage. The X- and Γ-profiles are sketched in inset (a) with their lowest energy quantum states. The current at high bias is plotted in inset (b) on a linear scale, together with the conductance. After Mendez *et al.* [22], with permission.

high bias is attributable to resonant tunnelling through the first quasi-bound state in the Γ-band and supports the present interpretation for the two structures at lower voltages.

Beresford *et al.* [24] used a more direct method of demonstrating Γ–X resonant intervalley tunnelling by using a GaAs/AlAs/GaAs single-barrier structure in which there is no quasi-bound state in the Γ-band so that no NDC characteristic is expected for Γ-band tunnelling. Nevertheless, NDC was clearly seen in the *I–V* characteristics for a 4-nm-thick AlAs single-barrier diode (see Fig. 3.12). This fact undoubtedly indi-

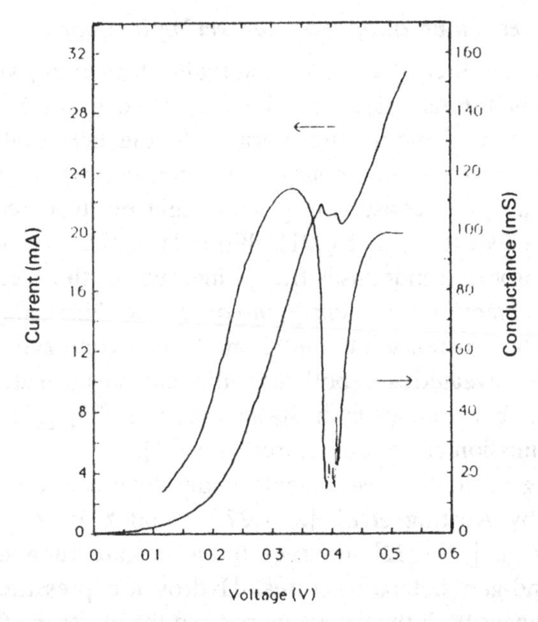

Figure 3.12 Current and conductance vs voltage for a 4-nm-thick GaAs/AlAs/GaAs single-barrier structure measured at 77 K. The sharp cut-off at 0.36 V is interpreted as a sign that the confined X state has passed below the conduction-band edge in the source electrode. After Beresford *et al.* [24], with permission.

cates the existence of resonant tunnelling through the confined state in the X-band. This is in quantitative agreement with theoretical work performed by Liu [21] in which a delta-function Γ–X-intervalley transition potential was introduced as an empirical parameter into the effective mass envelope function calculation.

Table 3.1. Pressure coefficients of direct and indirect energy gaps of $Al_xGa_{1-x}As$ at three energy points [33]

Points	Pressure coefficients (meV/kbar)
Γ-point dE_g^{Γ}/dP	$11.5 - 1.3x$
X-point dE_g^{X}/dP	-0.8
L-point dE_g^{L}/dP	2.8

3.2.2 Lateral momentum non-conserving intervalley tunnelling

Let us turn to another type of Γ–X-intervalley tunnelling which coexists with the resonant intervalley tunnelling described above. With regard to the conservation of momentum parallel to the heterointerfaces, this Γ–X-intervalley scattering induced by phonons, impurities or alloy potential is a $\mathbf{k}_{//}$-non-conserving process, and electrons can be transferred to the two (010) and (001) ellipsoids of the X-valley. In other words, this process randomises the momentum of the electrons and is thus called *lateral momentum non-conserving intervalley tunnelling* hereafter. The existence of lateral mode non-conserving intervalley tunnelling was revealed using both tunnelling current measurements with a high applied bias under hydrostatic pressure [25]–[30] and thermionic-field-emission current measurements [31].

Tunnelling current measurements under hydrostatic pressure were performed by Austing *et al.* [25]–[27], Mendez *et al.* [28]–[30] and Pritchard *et al.* [31], [32] in order to investigate tunnelling through indirect band-gap heterostructures. Hydrostatic pressure applied to AlGaAs changes both the valley energy and the electron effective mass. The pressure coefficients of the energy gaps, dE_g/dP, shown in Table 3.1, have a small dependence on the Al mole fraction, x, and $\Delta E(\Gamma_{GaAs} - X_{AlGaAs})$ is reduced at the rate of $-12.3\,\text{meV kbar}^{-1}$ while $\Delta E(\Gamma_{GaAs} - \Gamma_{AlGaAs})$ remains virtually unchanged. Thus a pressure-dependence measurement of the tunnelling current enables the ratio of the X-valley tunnelling to the Γ-valley tunnelling to be inferred.

Figure 3.13 shows the *I–V* characteristics for two AlAs/GaAs/AlAs double-barrier RTDs measured by Mendez *et al.* [30] under various pressures at 77 K: with an AlAs barrier thickness of (a) 2.3 nm and (b) with a thickness of 4.0 nm. While the peak current due to resonant tunnelling via the Γ-valley shows a weak dependence on pressure, the valley current in the higher-bias regime increases with increasing

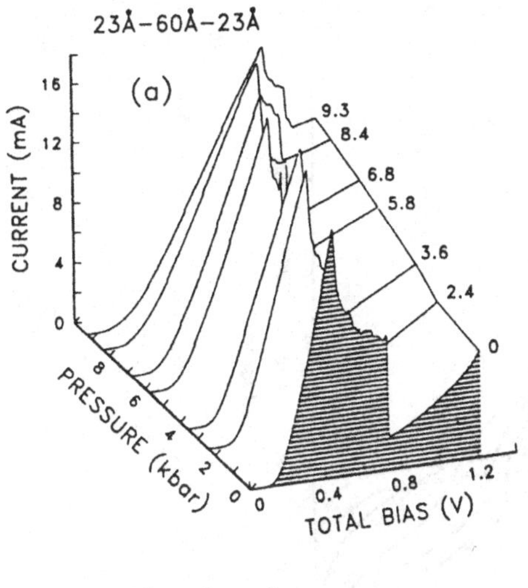

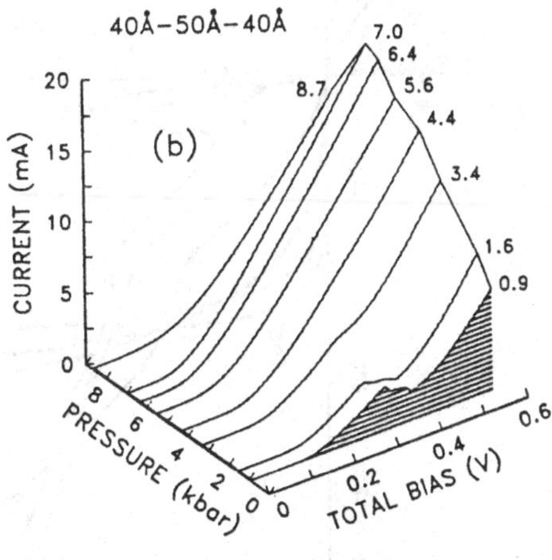

Figure 3.13 *I–V* characteristics of (a) 2.3 nm–6.0 nm–2.3 nm and (b) 4.0 nm–5.0 nm–4.0 nm AlAs/GaAs/AlAs heterostructures for representative pressures, at 77 K. After Mendez *et al.* [30], with permission.

pressure, resulting in a dramatic decrease in the *P/V* ratio. In the case of (b) in particular, the current peak disappears for pressure over 3.4 kbar. Mendez *et al.* [29] analysed the current increase in this high-applied-bias region using the Fowler–Nordheim formula [34]:

$$J = A\exp\left\{-\frac{4}{3}\sqrt{\left(\frac{2m_b^*}{\hbar^2}\right)}\frac{\Delta E(\Gamma_{\mathrm{GaAs}} - \mathrm{X}_{\mathrm{AlGaAs}})^{3/2}}{eE_b}\right\} \tag{3.18}$$

where E_b is an electric field applied to the tunnelling barrier. They found the effective mass of electrons in the barrier m_b^* to be about $0.26\,m_0$,

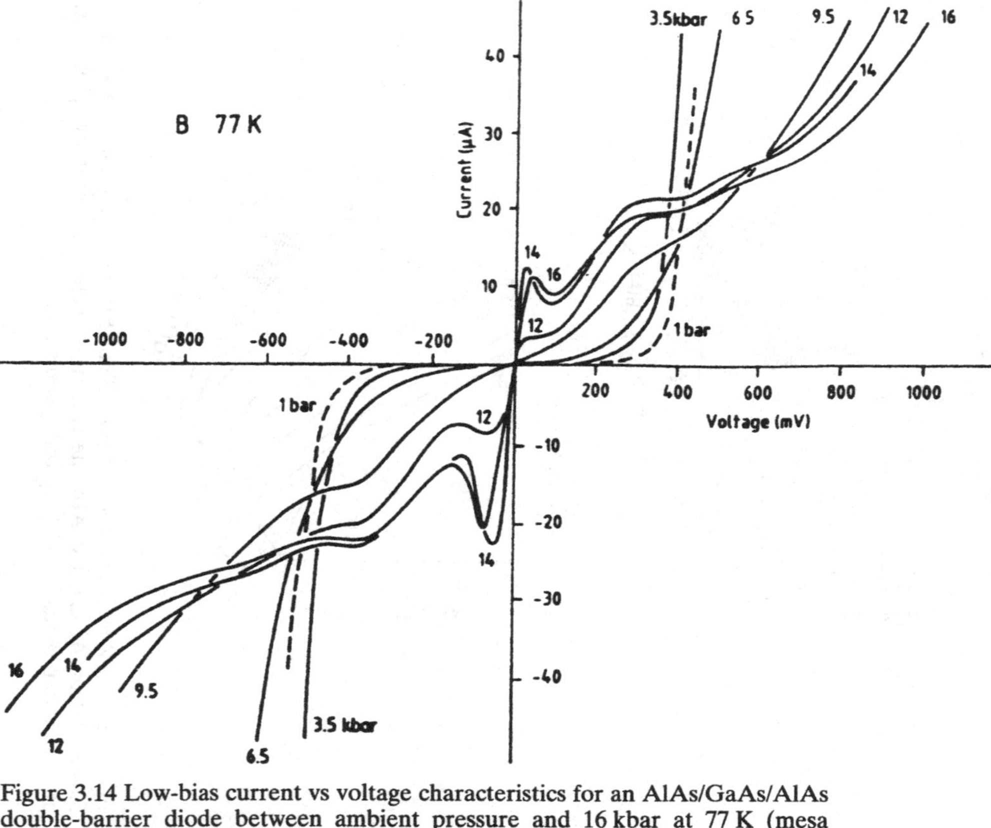

Figure 3.14 Low-bias current vs voltage characteristics for an AlAs/GaAs/AlAs double-barrier diode between ambient pressure and 16 kbar at 77 K (mesa diameter, 15 μm). After Austing *et al.* [26], with permission.

which is close to the transverse mass of electrons in the X-valley, m_t^*. This implies that the current in this regime also results from Γ–X-intervalley tunnelling which is, however, different from the resonant intervalley tunnelling studied earlier. While resonant intervalley tunnelling is a $\mathbf{k}_{/\!/}$-conserving process, in which the tunnelling electron has heavy longitudinal mass in the AlAs barriers, the present Γ–X-intervalley tunnelling is a $\mathbf{k}_{/\!/}$-non-conserving process which chooses the lighter effective mass, m_t^*, and achieves a relatively high transmission probability. This interpretation is consistent with the fact that a *P*/*V* current ratio remains visible in the high-pressure regime in the case of a thinner AlAs barrier structure (Fig. 3.13(a)), in contrast to the case of thicker barriers (Fig. 3.13(b)). This is because the lateral momentum non-conserving intervalley tunnelling contributes to the total current less when the thickness is reduced. A theoretical foundation for lateral momentum non-conserving intervalley tunnelling has not yet been reported, but Price [35] has demonstrated that the transverse X-valley mass is dominant for Fowler–Nordheim tunnelling through a single AlGaAs barrier when the electrons suffer from alloy disorder scattering.

Austing *et al.* [26] have observed resonant tunnelling via two transverse X-band quasi-bound states in AlAs/GaAs/AlAs double-barrier structures under high hydrostatic pressures. They measured the tunnelling current through a 4.2–7.2–4.0 nm double-barrier AlAs/GaAs/AlAs structure under a hydrostatic pressure up to 16 kbar and found a new resonance at very small biases above 12 kbar (see Fig. 3.14). As explained above, hydrostatic pressure applied to the device gives rise to a reduction of $\Delta E(\Gamma_{\mathrm{GaAs}} - \mathrm{X}_{\mathrm{AlAs}})$ and results in the band structure shown in Fig. 3.15 at 16 kbar. The lowest longitudinal and transverse quasi-bound states, E^{Xl} and E^{Xt}, formed in the X-band wells are located near the Γ-band edge in the GaAs regions, and both states may cause resonant tunnelling under an applied bias close to zero. Austing *et al.* have analysed the data in detail [26] by comparison with numerical calculations and have shown that the small peak seen near zero bias in Fig. 3.14 results from the resonant tunnelling between the transverse X-states, $E_1^{\mathrm{Xt}} \rightarrow E_2^{\mathrm{Xt}}$. A broad peak seen at about 300 mV has been attributed to associated LO-phonon-emission-assisted resonant tunnelling, $E_1^{\mathrm{Xt}} \rightarrow E_2^{\mathrm{Xt}} + \hbar\omega_{\mathrm{LO}}$.

A thermionic-field-emission study of lateral-momentum non-conserving Γ–X-intervalley tunnelling current has also been reported by Solomon *et al.* [36] for GaAs/$\mathrm{Al}_x\mathrm{Ga}_{1-x}$As/GaAs single-barrier structures

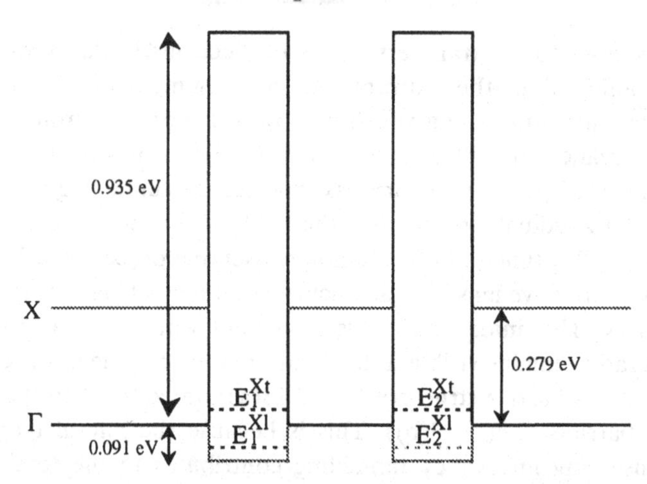

Figure 3.15 The Γ (solid lines) and X (grey lines) conduction-band edge profiles expected for an AlAs/GaAs/AlAs RTD under a pressure of 16 kbar. The lowest longitudinal and transverse X-point states are indicated in the barrier regions with E^{Xl} and E^{Xt}.

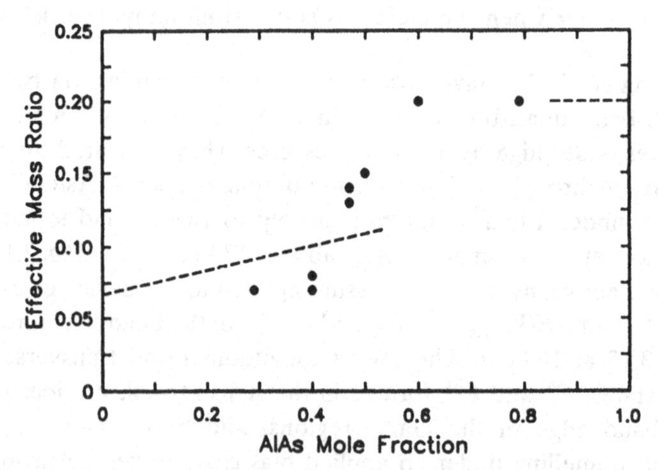

Figure 3.16 Electron effective mass m^* vs mole fraction, derived from the tunnelling data. The dashed lines are the Γ-effective mass (bottom) and transverse X effective mass (top) in AlGaAs. After Solomon *et al.* [36], with permission.

with an AlAs mole fraction, x, ranging from 0.3 to 0.8. They extracted the energy barrier height and effective mass involved in the tunnelling process. Figure 3.16 shows the mole fraction dependence of the effective mass, with a steep change being seen around $x = 0.5$. The effective mass obtained in the indirect gap region ($x > 0.5$) is 0.2 m_0, which is closer to the transverse X-mass and in agreement with the experimental results obtained from hydrostatic pressure measurements. These observations together indicate that lateral-momentum non-conserving Γ–X-intervalley tunnelling contributes greatly to the valley current in RTDs with indirect-gap structures. Structural parameters should be chosen carefully, taking account of these tunnelling processes in order to achieve large *P*/*V* current ratios.

3.3 Effects of impurity scattering in RTDs

Since an RTD consists of an undoped quantum well and barriers, impurity scattering is not supposed to have an influence on the resonant tunnelling processes. As a matter of fact, however, ionised donors in the adjacent contact regions, especially that below the resonant tunnelling structure, may diffuse into the barrier and well during the epitaxial growth process, giving rise to elastic Coulomb scattering centres. Experimental observation of impurity scattering was first reported by Goldman *et al.* [2] in the same $Al_{0.4}Ga_{0.6}As/GaAs/Al_{0.4}Ga_{0.6}As$ double-barrier RTD as that introduced in Section 3.1.1. They investigated the applied bias dependence of the magnetoconductance of the RTD in the valley current regime.

Figure 3.17 shows the *I*–*V* and d*I*/d*V*–*V* characteristics measured by Goldman *et al.* at 4.2 K under a magnetic field perpendicular to the barriers. The shallow peak in the *I*–*V* curve is the LO-phonon-emission-assisted satellite peak (see Section 3.1), and clear oscillations are seen in the d*I*/d*V*–*V* curve throughout the whole valley region. As explained in Section 3.1.3, a perpendicular magnetic field quantises the lateral electron motion into Landau levels. The oscillations indicate that some electrons experience inter-Landau-level transitions ($N' \neq N$) which are attributable to elastic impurity scattering which does not conserve lateral momentum.

As the perpendicular motion of electrons in a resonant tunnelling structure is quantised into resonant states, the selection rule for the possible final states of electrons after impurity scattering is different from that for free electrons. Theoretical studies of impurity scattering on resonant tunnelling have been reported by several groups [37]–[42]. Because lateral momentum is not conserved, a proper description of impurity scattering in RTDs requires 3D scattering theory. The transmission probability, *T*, is then a function of both the total electron energy and lateral momentum:

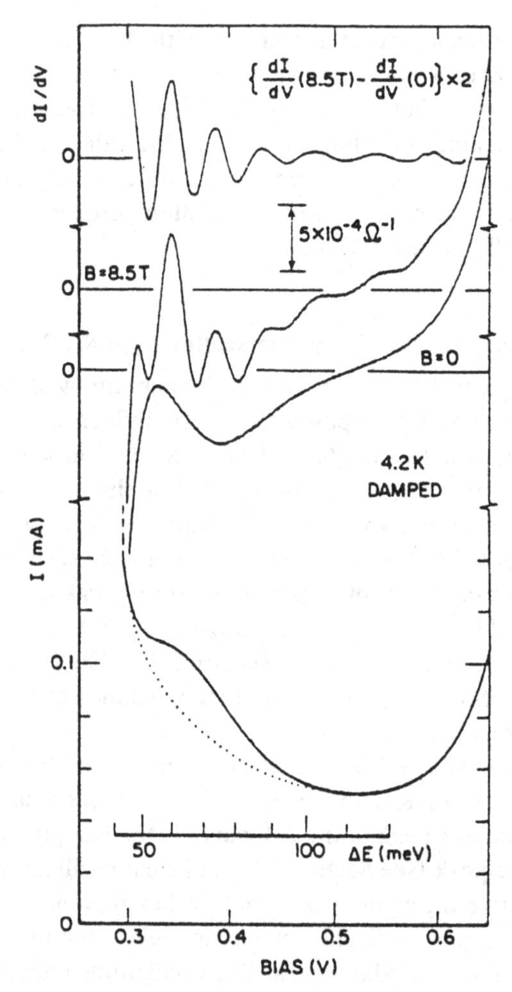

Figure 3.17 The *I–V* and d*I*/d*V* curves measured in the valley regime under a magnetic field perpendicular to the barriers. The top curve gives the difference between the differential conductivities at 8.5 and 0 T. The inset scale, ΔE, measures the calculated energy difference between the Fermi level in the emitter and the resonant level. After Goldman *et al.* [2], with permission.

$$T(\mathbf{k}^{in}_{/\!/},E) = \sum_{\mathbf{k}^{out}_{/\!/}} T(\mathbf{k}^{in}_{/\!/},\mathbf{k}^{out}_{/\!/},E) \tag{3.19}$$

where $\mathbf{k}^{in}_{/\!/},\mathbf{k}^{out}_{/\!/}$ are the lateral components of the wavenumbers of incoming and outgoing electrons and E is the total electron energy. Also, the transmission probability is dependent on the configuration of impurities $\{\mathbf{r}\} = \{\mathbf{r}_1,\mathbf{r}_2,\ldots,\mathbf{r}_N\}$, namely $T(\mathbf{k}^{in}_{/\!/},\mathbf{k}^{out}_{/\!/};E;\{\mathbf{r}\})$, although the tunnelling current that we observe is the configurational average over them all.

Many of the theoretical works reported so far have been based on 1D numerical calculations [37]–[40] which analyse the effect of impurity scattering on the diagonal component ($\mathbf{k}_{/\!/}^{\text{in}} = \mathbf{k}_{/\!/}^{\text{out}}$), neglecting the lateral-momentum non-conserving processes. Gu *et al.* [37] performed one-dimensional calculations for the effect of a few impurities in the quantum well on the transmission probability. For simplicity they adopted the delta-functional scattering potential rather than a realistic Coulomb potential. They calculated the transmission probability $\langle T \rangle$, which is a statistical average of T over all possible configurations of a few impurities in the well and is supposed to be a physically meaningful quantity. Figure 3.18(a)–(c) shows the averaged transmission probability of a double-barrier structure (3-nm-thick barriers and a 5.5-nm-thick well) around the first (Fig. 3.18(a)), second (Fig. 3.18(b)) and third (Fig. 3.18(c)) resonances. Lines 1 and 2 are the transmission probability of the double-barrier structure with no impurity and that averaged over the position of a single impurity in the well respectively. The single repulsive potential in the well pushes the resonant peak towards a higher energy, and the amount of shift depends on the impurity position. The resulting $\langle T \rangle$ is a superposition of these shifted peaks and thus has a broad transmission band centred at a higher energy than the original peak. This statistical broadening of the transmission peak obviously leads to a decrease in the current P/V ratio, as well as collisional broadening.

Now let us look at the effect of the lateral-momentum non-conserving elements ($\mathbf{k}_{/\!/}^{\text{in}} \neq \mathbf{k}_{/\!/}^{\text{out}}$). A fully 3D theoretical analysis has been reported by Fertig *et al.* [41], [42]. They used a microscopic perturbation theory, basically in the same manner as that used for LO-phonon scattering (Section 3.1.2). The theory is again based on the transfer Hamiltonian formula (Section 2.2.3), and the following term [41] is now introduced into the total Hamiltonian:

$$H_{\text{im}} = \sum_{\mathbf{k}_{/\!/}^{\text{w}},\mathbf{k}'^{\text{w}}_{/\!/}} U_{\text{im}}(\mathbf{k}'^{\text{w}}_{/\!/},\mathbf{k}^{\text{w}}_{/\!/})\, c^{+}_{\mathbf{k}'^{\text{w}}_{/\!/},k^{0}_{\perp}}\, c_{\mathbf{k}^{\text{w}}_{/\!/},k^{0}_{\perp}} \tag{3.20}$$

where U_{im} represents an effective interaction potential integrated over the positions of impurities in the quantum well, $\{\mathbf{r}_j\}$:

$$U_{\text{im}}(\mathbf{k}'_{/\!/},\mathbf{k}_{/\!/}) = a^2 V_0 \sum_j \mathrm{e}^{\mathrm{i}(\mathbf{k}'_{/\!/} - \mathbf{k}_{/\!/})\cdot\mathbf{r}_j} \tag{3.21}$$

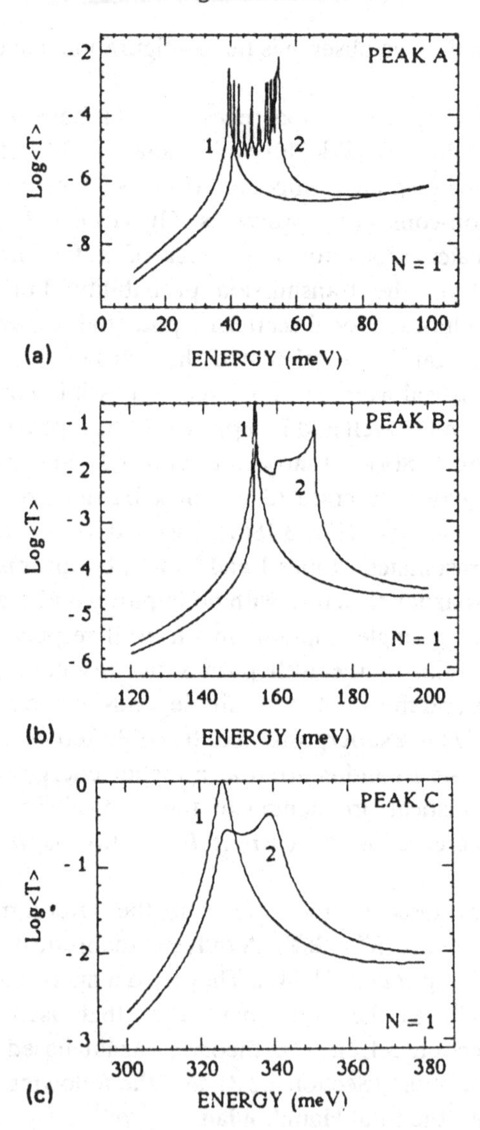

Figure 3.18 Logarithmic plots of averaged transmission probability over the position of scattering centres as a function of electron energy. Curve 1 corresponds to the case of no scattering and curve 2 to the presence of scattering: (a) for the first, (b) the second, and (c) the third resonances. After Gu *et al.* [37], with permission.

assuming the potential due to the *j*th impurity to be $a^2V_0\delta(\mathbf{r} - \mathbf{r}_j)$, where *a* is the size of the impurity. Fertig *et al.* calculated the zeroth- and first-order contributions to the transmission probability using a perturbation expansion in U_{im}. As we saw in Sections 2.2.3 and 3.1.2, the problem reduces to the calculation of the self-energy due to impurity scattering. The lowest-order self-energy is expressed simply as follows:

$$\Sigma_{im}(E) = \frac{a^2 V_0}{L^2} \sum_{\mathbf{k}_{/\!/}} \frac{1}{E - \epsilon'_{k_\perp^0} - \epsilon_{\mathbf{k}_{/\!/}} + i\Gamma} \tag{3.22}$$

The transmission probability to the lowest order, $T^{(0)}(\mathbf{k}_{/\!/}^{in},\mathbf{k}_{/\!/}^{out};E)$, is then found to be [42]

$$T^{(0)}(\mathbf{k}_{/\!/}^{in},\mathbf{k}_{/\!/}^{out};E) \cong \frac{\Gamma^2}{(E - \epsilon'_{k_\perp^0} - \epsilon_{\mathbf{k}_{/\!/}^{in}} - \mathrm{Re}\Sigma_{im}(E))^2 + \Gamma^2\left(1 + \dfrac{\mathrm{Im}\Sigma_{im}(E)}{\Gamma}\right)^2} \cdot\delta(\mathbf{k}_{/\!/}^{in} - \mathbf{k}_{/\!/}^{out}) \tag{3.23}$$

where the real and imaginary parts of the self-energy, eqn (3.22), are given by

$$\mathrm{Re}\Sigma_{im}(E) = \frac{m^* a^2 V_0}{4\pi} \ln\left\{\frac{(E - \epsilon'_{k_\perp^0} - \epsilon_{\Lambda_{/\!/}})^2 + \Gamma^2}{(E - \epsilon'_{k_\perp^0})^2 + \Gamma^2}\right\} \tag{3.24}$$

$$\mathrm{Im}\Sigma_{im}(E) = -\frac{m^* a^2 V_0}{2\pi}\left(\frac{\pi}{2} + \tan^{-1}\left\{\frac{(E - \epsilon'_{k_\perp^0})}{\Gamma}\right\}\right) \tag{3.25}$$

where $\epsilon_{\Lambda_{/\!/}}$ is the maximum for the lateral energy with a cut-off wave-number $\Lambda_{/\!/} \approx 1/a$. The real part (3.24) has a weak dependence on the electron energy and thus, in general, results in a shift of the resonant energy $\epsilon'_{k_\perp^0}$. The imaginary part gives rise to collisional broadening of a Lorentzian-type transmission probability with a slight asymmetry in the high-energy regime.

The first-order term, $T^{(1)}(\mathbf{k}_{/\!/}^{in},\mathbf{k}_{/\!/}^{out};E)$, is found to be [42]

$$T^{(1)}(\mathbf{k}_{/\!/}^{in},\mathbf{k}_{/\!/}^{out};E) \cong \frac{a^2 V_0}{L^2}\Gamma^2 \frac{1}{(E - \epsilon'_{k_\perp^0} - \epsilon_{\mathbf{k}_{/\!/}^{in}})^2 + \Gamma^2} \cdot \frac{1}{(E - \epsilon'_{k_\perp^0} - \epsilon_{\mathbf{k}_{/\!/}^{out}})^2 + \Gamma^2} \tag{3.26}$$

Fertig *et al.* have shown that $T^{(1)}(\mathbf{k}_{/\!/}^{in},\mathbf{k}_{/\!/}^{out};E)$ exhibits the following special selection rule for the scattering angle θ. Figure 3.19 shows an exit angle dependencies of $T^{(1)}(\mathbf{k}_{/\!/}^{in},\mathbf{k}_{/\!/}^{out};E)$ for two values of E: (a) below

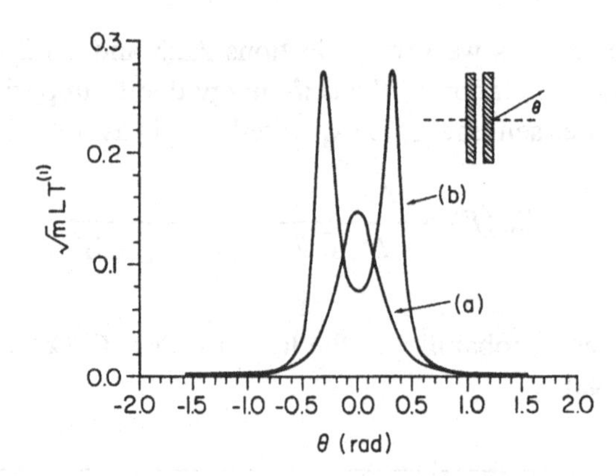

Figure 3.19 First-order correction to the transmission probability as a function of exit angle θ (see inset), for a 2D system, with $k_{/\!/}^{\rm in} = 0$: results for electron energy (a) below and (b) above the resonant energy. After Fertig *et al.* [42], with permission.

($E/\epsilon'_{k_\perp^0} = 0.9$) and (b) above ($E/\epsilon'_{k_\perp^0} = 1.1$) the resonant energy. The lateral momentum of the incident electron, $\mathbf{k}_{/\!/}^{\rm in}$, is set to be zero. For the case where $E > \epsilon'_{k_\perp^0}$, two sharp peaks are found at angles satisfying the condition $E - \hbar^2 \mathrm{k}_{/\!/}^{\rm out2}/2m^* = \epsilon'_{k_\perp^0}$. This result indicates that the electrons are favourably scattered on to a cone surface with a certain vertical angle: this is a *scattering induced focusing effect* [41],[42]. The total tunnelling current calculated from eqns (3.23) and (3.26) has been found to be in good agreement with that calculated using the phenomenological Breit–Wigner formula with an impurity-scattering-induced broadening parameter $\Gamma_{\rm im}$.

Experimental evidence of the above impurity-scattering-induced broadening has been observed in several ways. One method is to place impurities inside the quantum well of RTDs intentionally and to monitor a change in the current *P*/*V* ratio of the device. Wolak *et al.* [43] have produced four different $Al_{0.3}Ga_{0.7}As$(6 nm)/$Al_{0.3}Ga_{0.7}As$(6 nm) double-barrier RTDs with undoped, n-type (Si: 1×10^{18} cm^{-3}), p-type (Be: 1×10^{18} cm^{-3}) and compensated (alternating Si and Be) doping in the quantum well. Impurities are placed in the middle of the well over a distance of 1.67 nm. Apparently, both n- and p-type dopings cause additional space charge in the well, leading to different potential profiles at resonance. This fact makes quantitative comparison quite difficult, although the same potential distribution is expected for the case of compensated doping as for that of the undoped device. Figure 3.20 shows the *I*–*V* characteristics of these devices with an area of 2 μm

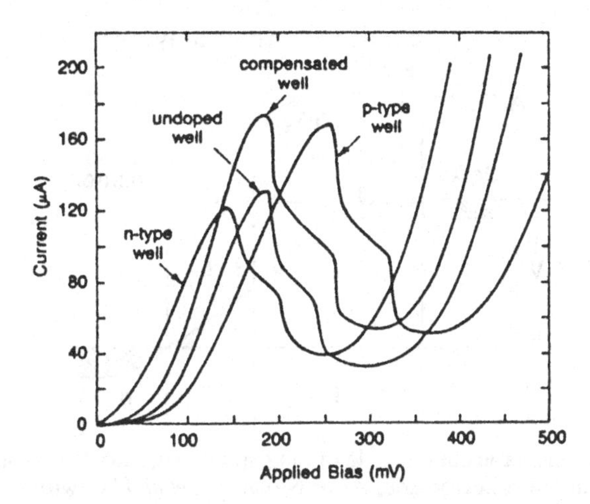

Figure 3.20 *I–V* characteristics measured at 77 K for samples with an n-doped well, an undoped well, a well with compensated doping, and a p-doped well. The peak to valley ratio is noticeably degraded in all the doped well samples. After Wolak *et al.* [43], with permission.

square measured at 77 K [43]. The n- and p-type samples have shifts in the peak voltage which agree with those estimated theoretically by taking account of the space charge effects on the potential profile [43]. It is found that, on the other hand, the sample with compensated doping has the resonant peak at almost the same voltage as the undoped well sample. Decreases in the *P*/*V* ratio are noticeable in all of the doped well samples and are attributable to impurity-scattering-induced broadening.

This experiment clearly demonstrates the effect of impurity scattering in a high doping concentration regime. However, we still have the question of how the background impurities in an undoped quantum well influence the RTD characteristics. The concentration of residual impurities is usually at least three orders of magnitude lower than the intentionally doped case. Nakagawa *et al.* [44] have succeeded in observing the effect of impurities diffused from a heavily doped contact region into a quantum well by using double-well RTDs (see Section 2.5). The double-well structure used for the experiment is shown in Fig. 3.21: the barriers and wells are $Al_{0.4}Ga_{0.6}As$ and GaAs, respectively, and the thickness of the spacer layer, *L*, between the third barrier and the heavily-doped n^+-GaAs collector region varies between 0 and 30 nm.

As shown in Section 2.5, much sharper current peaks are achieved in the double-well RTDs. Nakagawa *et al.* have monitored the change in

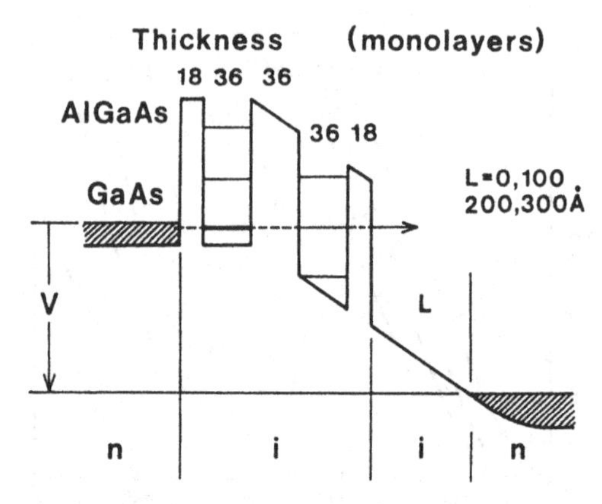

Figure 3.21 A schematic band profile of the triple-barrier diode with an undoped spacer layer on the collector side. After Nakagawa *et al.* [44], with permission.

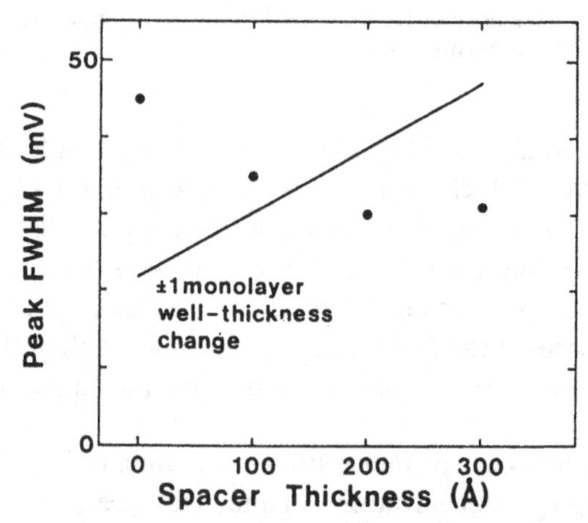

Figure 3.22 Dependence of the FWHM of the second resonance current peak on the spacer thickness. After Nakagawa *et al.* [44], with permission.

the width of the resonant current peaks with changing L. The measured full width at half maximums (FWHMs) of the current peak under the resonance condition illustrated in Fig. 3.21 are plotted as solid circles in Fig. 3.22 as a function of L. The FWHM decreases gradually as L increases and appears to reach a constant value with L above 20 nm. At the same time the current P/V ratio is found to increase with increasing L. This implies that ionised Si atoms, diffused from the contact region, cause visible collisional broadening in these structures which can be avoided by introducing a spacer layer thicker than 20 nm. In Fig. 3.22 a solid line is also drawn to show the theoretically estimated statistical broadening of the current peak due to well–barrier interface roughness with a deviation of one monolayer in thickness (*interface-roughness-induced broadening*). This indicates that collisional broadening due to diffused impurities is in general as small as roughness-induced broadening.

3.4 References

[1] C. J. Goodings, Variable-area resonant tunnelling diodes using implanted gates, PhD thesis, Cambridge University, 1993.

[2] V. J. Goldman, D. C. Tsui and J. E. Cunningham, Evidence for LO-phonon-emission-assisted tunneling in double-barrier heterostructures, *Phys. Rev.*, **B36**, 7635, 1987.

[3] F. Chevoir and B. Vintor, Calculation of phonon-assisted tunneling and valley current in a double-barrier diode, *Appl. Phys. Lett.*, **55**, 1859, 1989.

[4] F. Chevoir and B. Vintor, Resonant and scattering-assisted magnetotunneling, in *Resonant Tunneling in Semiconductors: Physics and applications*, edited by L. L. Chang, E. E. Mendez and C. Tejedor (Plenum, New York, 1990).

[5] S. Das Sarma and B. A. Mason, Optical phonon interaction effects in layered semiconductor structures, *Ann. Phys.*, **163**, 78, 1985.

[6] N. S. Wingreen, K. W. Jacobsen and J. W. Wilkins, Resonant tunneling with electron–phonon interaction: An exactly solvable model, *Phys. Rev. Lett.*, **61**, 1396, 1988.

[7] M. Jonson, Quantum-mechanical resonant tunneling in the presence of a boson field, *Phys. Rev.*, **B39**, 5924, 1989.

[8] W. C. T. F. Zheng, P. Hu, B. Yudanin and M. Lax, Model of phonon-assisted electron tunneling through a semiconductor double barrier, *Phys. Rev. Lett.*, **63**, 418, 1989.

[9] P. J. Turley and S. W. Teitsworth, Electronic wave function and electron-confined-phonon matrix elements in $GaAs/Al_xGa_{1-x}As$ double-barrier resonant-tunneling structures, *Phys. Rev.*, **B44**, 3199, 1991.

[10] P. J. Turley and S. W. Teitsworth, Phonon-assisted tunneling due to localized modes in double-barrier structures, *Phys. Rev.*, **B44**, 8181, 1991.

[11] N. Mori, K. Taniguchi and C. Hamaguchi, Effects of electron–interface–phonon interaction on resonant tunnelling in double-barrier heterostructures, *Semicond. Sci. Technol.*, **7**, B83, 1992.

[12] N. Zou and K. A. Chao, Inelastic electron resonant tunneling through a double-barrier nanostructure, *Phys. Rev. Lett.*, **69**, 3224, 1992.

[13] A. A. Klochikhin, Spectrum of a free polaron and convergence radius in perturbation theory, *Sov. Phys.-Solid State*, **21**, 1770, 1980.

[14] M. L. Leadbeater, E. S. Alves, L. Eaves, M. Henini, O. H. Hughes, A. Celeste, J. S. Portal, G. Hill and M. A. Pate, Magnetic field studies of elastic scattering and optic-phonon emission in resonant-tunneling devices, *Phys. Rev.*, **B39**, 3438, 1989.

[15] C. J. Goodings, H. Mizuta and J. R. A. Cleaver, Electrical studies of charge build-up and phonon-assisted tunnelling in double-barrier materials with very thick spacer layers, *J. Appl. Phys.*, **75**, 363, 1994.

[16] E. E. Mendez, E. Calleja, C. E. T. Goncalves da Silva L. L. Chang and W. I. Wang, Observation by resonant tunneling of high-energy states in GaAs–$Ga_{1-x}Al_xAs$ quantum wells, *Phys. Rev.*, **B33**, 7368, 1968.
[17] H. Akera, S. Wakahara and T. Ando, Connection rule of envelope functions at heterostructures, *Surface Sci.*, **196**, 694, 1988.
[18] D. Z. Y. Ting, M. K. Jackson, D. H. Chow, J. R. Söderström, D. A. Collins and T. C. McGill, X-point tunneling in AlAs/GaAs double barrier heterostructures, *Solid State Electron.*, **32**, 1513, 1989.
[19] J. N. Schulman and Y-C. Chang, Band mixing in semiconductor superlattices, *Phys. Rev.*, **B31**, 2056, 1984.
[20] N. A. Cade, S. H. Parmar, N. R. Couch and M. J. Kelly, Indirect resonant tunnelling GaAs/AlAs, *Solid State Communications*, **64**, 283, 1990.
[21] H. C. Liu, Resonant tunneling through single layer heterostructures, *Appl. Phys. Lett.*, **51**, 1019, 1987.
[22] E. E. Mendez, W. I. Wang, E. Calleja and C. E. T. Gonçalves da Silva, Resonant tunneling via X-point states in AlAs-GaAs-AlAs heterostructures, *Appl. Phys. Lett.*, **50**, 1263, 1987.
[23] A. R. Bonnefoi, T. C. McGill, R. D. Burnham and G. R. Anderson, Observation of resonant tunneling through GaAs quantum well states confined by AlAs X-point barriers, *Appl. Phys. Lett.*, **50**, 344, 1987.
[24] R. Beresford, L. F. Luo, W. I. Wang and E. E. Mendez, Resonant tunneling through X-valley states in GaAs/AlAs/GaAs single barrier heterostructures, *Appl. Phys. Lett.*, **55**, 1555, 1989.
[25] D. G. Austing, P. C. Klipstein, J. S. Roberts and G. Hill, Resonant tunnelling between X-levels in a GaAs/AlAs/GaAs/AlAs/GaAs device above 13 kbar, *Solid State Communications*, **75**, 697, 1990.
[26] D. G. Austing, P. C. Klipstein, A. W. Higgs, H. J. Hutchinson, G. W. Smith, J. S. Roberts and G. Hill, X- and Γ-related tunneling resonances in GaAs/AlAs double-barrier structures at high pressures, *Phys. Rev.*, **B47**, 1419, 1993.
[27] D. G. Austing, P. C. Klipstein, J. S. Roberts, C. B. Button and G. Hill, Tunneling resonances at high pressure in double-barrier structures with AlAs barriers thicker than 50 Å, *Phys. Rev.*, **B48**, 11905, 1993.
[28] E. E. Mendez and L. L. Chang, Tunneling between two dimensional electron gases, *Surface Science*, **229**, 173, 1990.
[29] E. E. Mendez, E. Calleja and W. I. Wang, Tunneling through indirect-gap semiconductor barriers, *Phys. Rev.*, **B34**, 6026, 1986.
[30] E. E. Mendez, E. Calleja and W. I. Wang, Inelastic tunneling in AlAs–GaAs–AlAs heterostructures, *Appl. Phys. Lett.*, **53**, 977, 1988.
[31] R. Pritchard, P. C. Klipstein, N. R. Couch, T. M. Kerr, J. S. Roberts, P. Mistry, B. Soylu and W. M. Stobbs, High-pressure studies of resonant tunnelling in a graded parameter superlattice and in double barrier structures of GaAs/AlAs, *Semicond. Sci. Technol.*, **4**, 754, 1989.
[32] R. Pritchard, D. G. Austing, P. C. Klipstein, J. S. Roberts, A. W. Higgs and G. W. Smith, The suppression by pressure of negative differential resistance in GaAs/GaAlAs double barrier structures, *J. Appl. Phys.*, **68**, 205, 1990.
[33] S. Adachi, GaAs, AlAs, and $Al_xGa_{1-x}As$: Material parameters for use in research and device applications, *J. Appl. Phys.*, **58**, R1, 1985.
[34] E. L. Murphy and R. H. Good, Jr, Thermionic emission, field emission, and the transition region, *Phys. Rev.*, **102**, 1464, 1956.

[35] P. J. Price, Tunneling in AlGaAs by Γ–X scattering, *Surface Science*, **196**, 394, 1988.
[36] P. M. Solomon, S. L. Wright and C. Lanza, Perpendicular transport across (Al,Ga)As and the Γ to X transition, *Superlattices and Microstructures*, **2**, 521, 1986.
[37] B. Gu, C. Coluzza, M. Mangiantini and A. Frova, Scattering effects on resonant tunneling in double-barrier heterostructures, *J. Appl. Phys.*, **65**, 3510, 1989.
[38] A. F. M. Anwar, R. B. LaComb and M. Cahay, Influence of impurity scattering on the traversal time and current–voltage characteristics of resonant tunneling structures, *Superlattices and Microstructures*, **11**, 131, 1992.
[39] C. J. Arsenault and M. Meunier, Proposed new resonant tunneling structures with impurity planes of deep levels in barriers, *J. Appl. Phys.*, **66**, 4305, 1989.
[40] Jin-feng Zhang and Ben-yuan Gu, Effects of a localized state inside the barrier on the temporal characteristics of electron tunneling in double-barrier quantum wells, *Phys. Rev.*, **B44**, 8204, 1991.
[41] H. A. Fertig and S. Das. Sarma, Elastic scattering in resonant tunneling systems, *Phys. Rev.*, **B40**, 7410, 1989.
[42] H. A. Fertig, Song He and S. Das. Sarma, Elastic-scattering effects on resonant tunneling in double-barrier quantum-well structures, *Phys. Rev.*, **B41**, 3596, 1990.
[43] E. Wolak, K. L. Lear, P. M. Pinter, E. S. Hellman, B. G. Park, T. Well, J. S. Harris, Jr and D. Thomas, Elastic scattering centers in resonant tunneling diodes, *Appl. Phys. Lett.*, **53**, 201, 1988.
[44] T. Nakagawa, T. Fujita, Y. Matsumoto, T. Kojima and K. Ohta, Sharp resonance characteristics in triple-barrier diodes with a thin undoped spacer layer, *Jpn. J. Appl. Phys.*, **26**, L980, 1987.

4

Femtosecond dynamics and non-equilibrium distribution of electrons in resonant tunnelling diodes

Following the study in Chapter 3 of the effects of elastic and inelastic scattering on the transmission probability function, this chapter investigates *non-equilibrium electron distribution* in RTDs. Electron distribution in the triangular potential well in the emitter is studied first (Section 4.1). Then dissipative quantum transport theory is presented based on the Liouville–von-Neumann equation for the statistical density matrix (Section 4.2.1). Numerical calculations are carried out in order to analyse the *femtosecond dynamics of the electrons* (Section 4.2.2) and the dynamical *space charge build-up* in the double-barrier structure which gives rise to the intrinsic current bistability in the NDC region (Section 4.3.1). Next experimental studies of the charge build-up phenomenon are presented using magnetoconductance measurements (Section 4.3.2) and photoluminescence measurements (Section 4.3.3). Finally, the effects of magnetic fields on intrinsic current bistability are studied (Section 4.4).

4.1 Non-equilibrium electron distribution in RTDs

Let us start with a discussion on electron distribution in the emitter. We have seen in Section 2.4 that the electronic states in the emitter become 2D in the pseudo-triangular potential well formed between the thick spacer layer and the tunnelling barrier (see Fig. 2.16). Sharper current peaks observed for Materials 2 and 3 (Fig. 2.18) have been attributed to the 2D–2D nature of resonant tunnelling. This interpretation is based upon an assumption that the electrons in the triangular well are well thermalised, and that local equilibrium is achieved. This is not necessarily the case as the incoming electrons are redistributed in this region between both quasi-bound (2D) states and scattering (3D) states. As

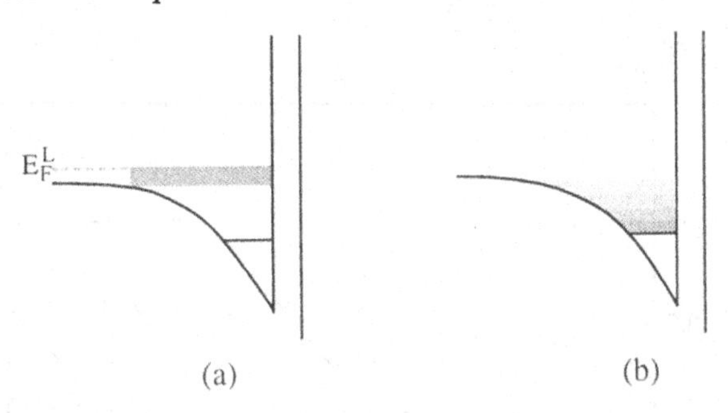

Figure 4.1 Two extreme cases for electron accumulation in a pseudo-triangular potential well: (a) no charge build-up due to ballistic electron transport, and (b) significant build-up caused by electrons which undergo energy dissipation processes.

long as the transport is completely ballistic, as is assumed in the global coherent tunnelling model, the incoming electrons pass over this triangular well without being thermalised into the quasi-bound states which are located below the conduction-band edge of the emitter (see Fig. 4.1(a)). This, however, is not a realistic assumption since the incoming electrons may undergo dissipative scattering processes and become redistributed on to these quasi-bound states (see Fig. 4.1(b)). In this case, the electron distribution is not expressed simply by the Fermi–Dirac distribution function (which basically applies to systems in equilibrium) but by a complicated *non-equilibrium* distribution function. Thus the electrons tunnelling into the resonant state in the double-barrier structure may come from both the 2D quasi-bound states (2D–2D RT) and the 3D scattering states (3D–2D RT). If electrons from both locations contribute equally to resonant tunnelling, some distinguishable structures might be seen in the *I–V* characteristics.

The current–voltage and conductance–voltage characteristics are given for Materials 2 and 3 at various temperatures in Figs. 4.2 and 4.3 [1]. A single sharp current peak is seen at low temperatures which is thought to reflect the nature of the 2D–2D resonant tunnelling discussed earlier. For both materials, however, a broad current shoulder becomes noticeable below the main resonance at higher temperatures. Such a feature has also been observed by Gobato *et al.* [2] and Zheng *et al.* [3], who all saw similar characteristics and ascribed them to the electrons which pass through the 3D scattering states ballistically. The contributions of these 3D ballistic electrons have been modelled successfully by Chevoir and Vinter [4]. Also, Goldman *et al.* [5] have discussed a more

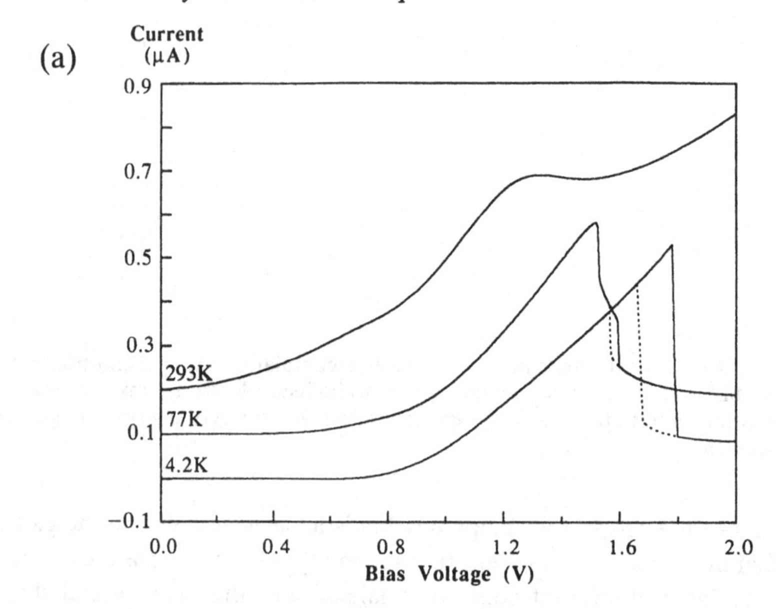

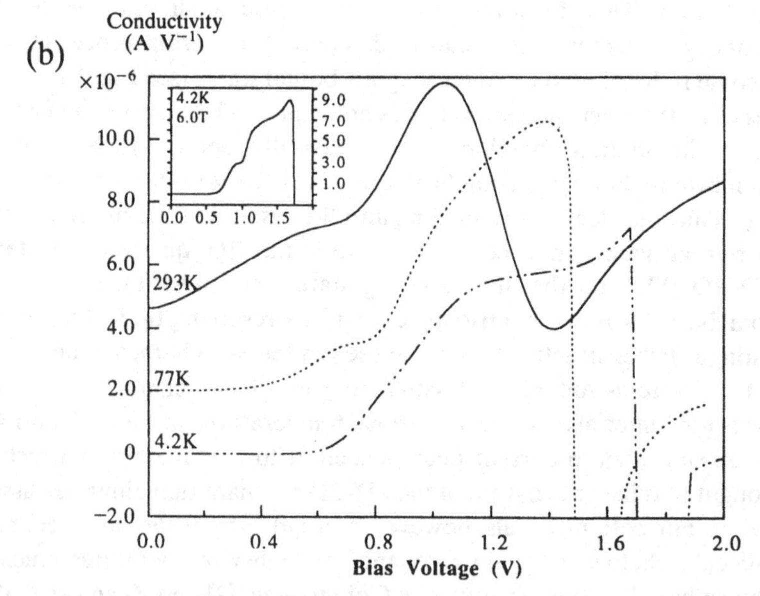

Figure 4.2 Current (a) and conductance (b) versus voltage measured for Material 2 at various temperatures. For clarity, the curves have been offset in both graphs. Dotted lines in (a) show hysteresis in the NDC regime. The inset in (b) shows the effect of a 6.0 T magnetic field perpendicular to the barriers on conductance at 4.2 K (for a smaller device). After Goodings [1], with permission.

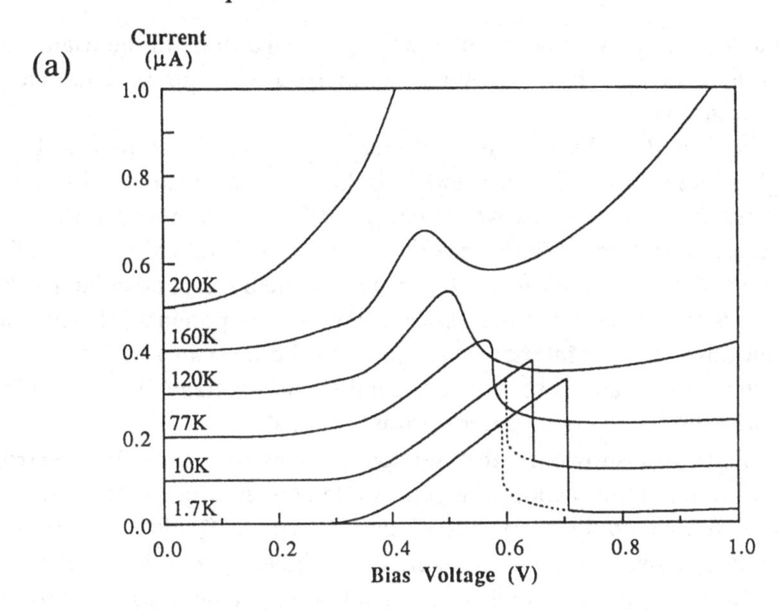

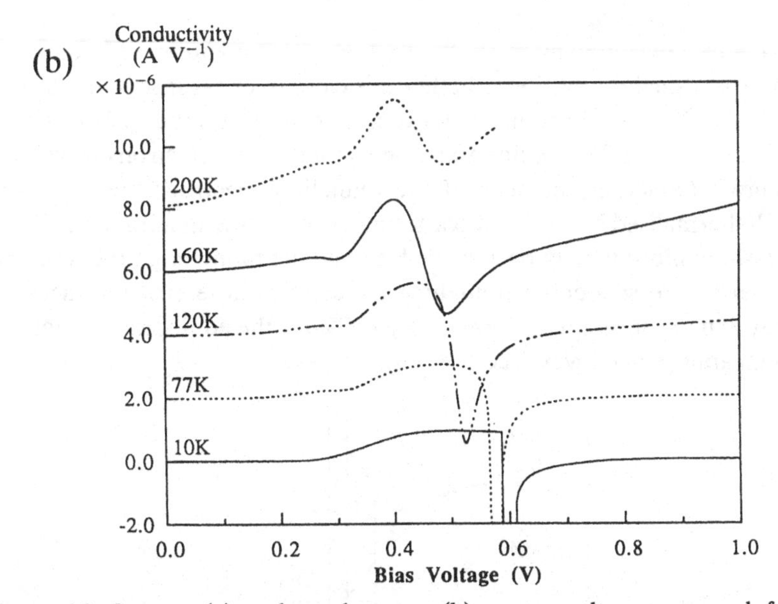

Figure 4.3 Current (a) and conductance (b) versus voltage measured for Material 3 at various temperatures. For clarity, the curves have been offset in both graphs. Dotted lines in (a) show hysteresis in the NDC regime. After Goodings [1], with permission.

complex case in which only the lowest quasi-eigenstate in the triangular well is resolved, while the others are merged with the 3D continuous scattering states.

The fact that the current shoulders are observed more clearly at higher temperatures is consistent with this theoretical explanation since the number of electrons which occupy the 3D continuous states will increase with temperature. In addition, these theoretical ideas readily predict that the application of a magnetic field perpendicular to the barriers should separate the 3D from the 2D components [3]. This has been observed in Material 2, as shown in the inset in Fig. 4.2(b): two distinct peaks can be seen in the conductance–voltage characteristics, even at the relatively high temperature of 4.2 K.

The results shown in this section demonstrate that the electron distribution in the pseudo-triangular well in the emitter is not expressed by a simple equilibrium distribution function. Reality lies in the situation between the two extreme cases shown in Fig. 4.1(a) and (b). Electrons in the non-equilibrium distribution are able to tunnel into the double-barrier structure, making the distribution between the double barriers more complex.

Let us move on to electron distribution in the quantum well, for which there is a similar question: whether it is close to or far from equilibrium. The distribution function in the quantum well may be very different from the Fermi distribution function as long as the electrons travel in a ballistic (coherent) manner. If the tunnelling electrons are, however, well thermalised by frequent scattering events in the quantum well, local quasi-equilibrium may be achieved, and so the tunnelling process draws closer to the sequential tunnelling picture. In these circumstances we may define a local quasi-Fermi energy, E_F^W, for the equilibrated electrons in an approximate way (see Fig. 4.4).

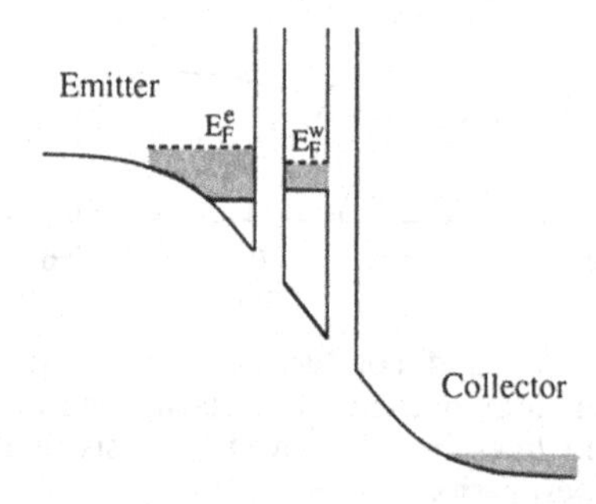

Figure 4.4 Schematic energy-band diagram and quasi-Fermi levels in the emitter and quantum well at resonance.

The true situation in the quantum well really depends on the materials being used. If the electron dwell time is far longer than the typical scattering time, as is the case for structures with relatively thick barriers, the above quasi-equilibrium assumption may be true. This point has been the subject of lively discussion, and is related to other questions such as whether coherent or sequential tunnelling is the more appropriate picture, and whether the current bistability of RTDs results from a dynamic space charge build-up in the quantum well or not. There have been many experimental and theoretical investigations on these issues. Theoretical analysis should employ dissipative quantum transport theory, which is very difficult to handle and requires an enormous amount of numerical calculation. Some numerical results, however, have been obtained successfully recently, based on the self-consistent analysis of the quantum Liouville equation. In experiments, direct observation of space charge build-up has been achieved by using a variety of techniques such as magnetoconductance and capacitance measurements, photoluminescence intensity and lineshape analysis and electroluminescence intensity analysis. These theoretical and experimental studies of non-equilibrium electron distribution will be presented in detail in the pages to come.

4.2 Theory of dissipative quantum transport in RTDs

This section presents a numerical study of dissipative tunnelling in RTDs based upon the Liouville–von-Neumann equation for the density matrix [6], which is the most fundamental equation in statistical quantum mechanical physics. The density matrix is a correlation function, and its off-diagonal elements directly measure the phase-coherence of the wavefunctions: one of the most significant parameters in electron-wave device design. In Section 4.2.1 the density matrix is first calculated in thermal equilibrium by using a complete set of scattering states which are obtained through self-consistent calculations of Schrödinger's and Poisson's equations. Using the thermal equilibrium density matrix as an initial state, the time-dependent density matrix equation is then solved directly in Section 4.2.2. The Hartree self-consistent field model and the relaxation-time approximation are introduced for electron–electron interactions and scattering processes respectively.

4.2.1 The time-dependent statistical density matrix and the Wigner distribution function

The first step in the simulation is to find the correct *statistical density matrix* for a system in thermal equilibrium, which is used as an initial condition for the time-dependent Liouville equation. Provided that an unperturbed quantum mechanical open system is in thermal equilibrium with its surroundings, the statistical density matrix based on an independent-particle approximation is expressed by the following equation:

$$\rho_0(\mathbf{x},\mathbf{x}';0) = \langle \Psi(\mathbf{x})\Psi^*(\mathbf{x}')\rangle = \sum_{\mathbf{k},\sigma} \Psi_{\mathbf{k}}(\mathbf{x})\Psi_{\mathbf{k}}^*(\mathbf{x}')f(\mathbf{k}) \tag{4.1}$$

where $\Psi_{\mathbf{k}}(\mathbf{x})$ is the scattering state of the unperturbed Hamiltonian (see eqn (2.9)), and $f(\mathbf{k})$ is the Fermi–Dirac distribution function. Equation (4.1) shows that the statistical density matrix is a function of the two positions $\mathbf{x}$ and $\mathbf{x}'$: the diagonal elements represent the real electron density in the system and the off-diagonal elements measure the phase-correlations of the wavefunctions. Also, it is sometimes convenient to introduce the following *Wigner distribution function*, $f_W(\chi,\mathbf{k},t)$, which is the Weyl transform of the density matrix:

$$f_W(\chi,\mathbf{k},t) = \int \frac{d\zeta}{2\pi} e^{i\mathbf{k}\cdot\zeta}\rho(\chi,\zeta,t) \tag{4.2}$$

where χ and ζ are the absolute and relative coordinates, $(\mathbf{x} + \mathbf{x}')/2$ and $\mathbf{x} - \mathbf{x}'$ respectively. This is analogous to the classical distribution function defined in the $(\mathbf{k},\mathbf{x})$ phase space. A complete set of scattering states required to calculate the density matrix ((4.1)) is obtained through a self-consistent calculation similar to that used for Schrödinger's equation, eqn (2.9), and Poisson's equation, eqn (2.28) (see Section 2.2.2).

For the present purpose, however, Schrödinger's equation is solved using the finite-difference method rather than the transfer matrix method. From the point of view of numerical calculations, the transfer matrix method studied in Chapter 2 is more accurate than the finite-difference method because the wavefunctions between the mesh points are stored as plane-wave states. However, the use of the solutions obtained by the transfer matrix method sometimes results in an unrealistic current density at heterointerfaces when the resulting density matrix is used as an initial solution for the time-dependent Liouville equation.

This is because of the difference in the ways in which the Schrödinger equation and the density matrix equation are discretised. In the following description we adopt precisely the same finite-difference discretisation for both equations.

A set of incoming scattering states (emitter- or collector-incident plane waves) is again used as the boundary condition to solve the Schrödinger equation. However, when the finite-difference method is used to solve the Schrödinger equation, the solutions are no longer expressed as coefficients of plane waves. It is therefore necessary to decompose the wavefunctions obtained at the edge of the modelled region into incident and reflected plane waves in order to find a normalisation factor. From the unnormalised solutions for the first two mesh points, we evaluate the coefficients for the incident and reflected waves and normalise the wave functions. The accuracy of this normalisation method depends on the first mesh-spacing and, to some extent, can be judged from the resultant electron density calculated at the emitter and collector edges.

The equilibrium density matrix has been calculated for an AlGaAs/GaAs/AlGaAs double-barrier structure which consists of an undoped GaAs quantum well 5 nm thick, two undoped $Al_{0.33}Ga_{0.67}As$ barriers 4 nm thick, two undoped GaAs spacer layers 6 nm thick and highly doped n-type GaAs ($N_D = 1 \times 10^{18}\ cm^{-3}$) emitter and collector layers 20 nm thick. Figure 4.5(a) shows the real parts of the equilibrium density matrix (the imaginary parts are zero), and Fig. 4.5(b) shows the energy-band diagram calculated self-consistently at a temperature of 300 K. An energy mesh with a spacing of 0.5 meV was used to calculate and store all the wavefunctions of the system with eigenenergies up to 200 meV. The normalisation method for the wavefunctions (described above) appears to be successful since charge neutral regions, assured by the flat energy band, are obtained near the emitter and collector edges. The small peak in the centre of Fig. 4.5(a) represents electrons accumulated in the quantum well. These electrons are thermally distributed around the first quasi-bound state (the resonant state) at an energy of about 60 meV measured from the bottom of the quantum well. The steep decrease in electron density near the double barriers stems from a quantum repulsion of electron waves [7].

A typical grid adopted for the density matrix calculation of a resonant tunnelling diode is shown in Fig. 4.6: the mesh-spacing is chosen to be smaller in the region of the double-barrier structure. This sort of non-uniform mesh is used to save computational time and memory when

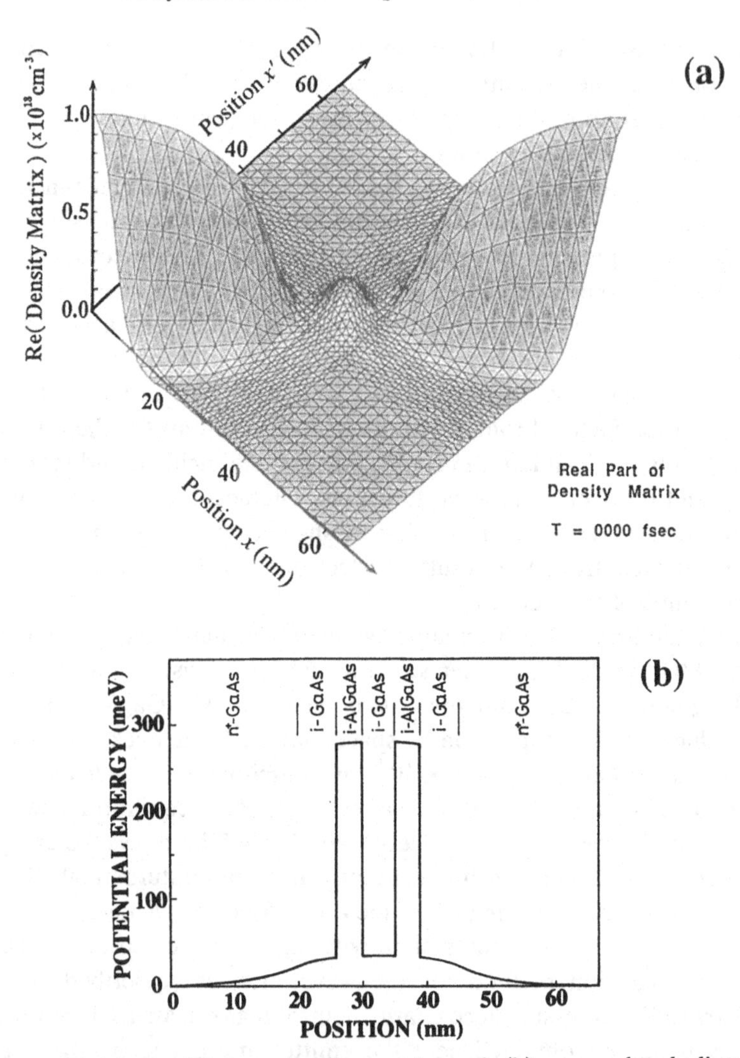

Figure 4.5 (a) Equilibrium density matrix and (b) energy-band diagram, calculated self-consistently for the AlGaAs/GaAs double-barrier RTD at a temperature of 300 K. The transverse axis in (b) has been adjusted so that the diagonal of the density matrix, the electron distribution, is displayed on the same scale.

solving the Liouville equation, although it complicates the coefficient matrix of the discretised equation.

As mentioned earlier, the off-diagonal components of the density matrix measure the phase-correlations between the electron wavefunctions at different positions. In Fig. 4.5(a), the density matrix in the

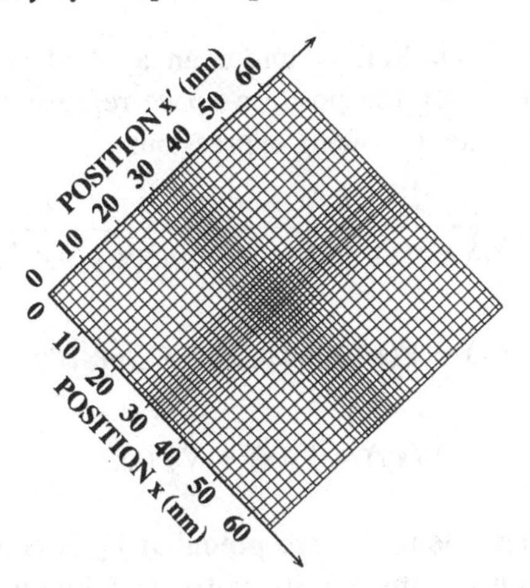

Figure 4.6 Positional grid used for the density matrix calculation of a double-barrier RTD. The mesh spacing has been chosen to be small for AlGaAs barriers and the GaAs quantum well.

emitter and collector regions varies steeply along the cross-diagonal direction, and its dependence on the relative distance, $\mathbf{x}-\mathbf{x}'$, is expressed approximately by the following a Gaussian-like analytical form for the free electron density matrix:

$$\rho_{\mathrm{FE}}(\mathbf{x},\mathbf{x}') \cong \frac{1}{\sqrt{(2\pi^3)}\lambda_{\mathrm{T}}^3}\{\exp - (\mathbf{x}-\mathbf{x}')^2/2\lambda_{\mathrm{T}}^2 + E_{\mathrm{F}}/k_{\mathrm{B}}T\} \qquad (4.3)$$

where the thermal coherence length of the electron wave, λ_{T}, is defined as follows:

$$\lambda_{\mathrm{T}}^2 = \hbar^2/(m^* k_{\mathrm{B}} T) \qquad (4.4)$$

The thermal coherence length, which is about 7 nm for GaAs at room temperature, is a critical length beyond which the correlation between thermally distributed free electrons decreases steeply because of the superposition of electron waves with various phases.

The second step in the simulation is to solve the time-dependent Liouville–von-Neumann equation for the density matrix. The equilibrium density matrix, ρ_0, obtained above is used as an initial state for the density matrix equation. The time evolution of the density matrix under an applied electric field is governed by

$$\frac{\partial \rho}{\partial t} = \frac{1}{i\hbar}[H(t),\rho] + C_{\mathrm{col}}\rho \qquad (4.5)$$

where $H(t)$ is the Hamiltonian under an applied bias, and C_{col} is a collision operator. Using the position–space representation, the above equation is given by the following expression:

$$\frac{\partial\rho(\mathbf{x},\mathbf{x}';t)}{\partial t} = \frac{i\hbar}{2}\left[\frac{\partial}{\partial\mathbf{x}}\left(\frac{1}{m^*(\mathbf{x})}\frac{\partial}{\partial\mathbf{x}}\right) - \frac{\partial}{\partial\mathbf{x}'}\left(\frac{1}{m^*(\mathbf{x}')}\frac{\partial}{\partial\mathbf{x}'}\right)\right]\rho(\mathbf{x},\mathbf{x}';t)$$

$$+ \frac{1}{i\hbar}\{V(\mathbf{x};t) - V(\mathbf{x}';t)\}\rho(\mathbf{x},\mathbf{x}';t) + C_{col}\rho(\mathbf{x},\mathbf{x}';t) \tag{4.6}$$

$$V(\mathbf{x};t) = V_0(\mathbf{x}) + V_{SC}(\mathbf{x};t) \tag{4.7}$$

where the time-dependent Hartree potential $V_{SC}(\mathbf{x};t)$ is determined by the diagonal elements of the density matrix as follows:

$$\frac{d}{d\mathbf{x}}\left\{\epsilon(\mathbf{x})\frac{dV_{SC}(\mathbf{x};t)}{d\mathbf{x}}\right\} = -e\{N_D^+(\mathbf{x}) - \mathrm{Re}(\rho(\mathbf{x},\mathbf{x};t))\} \tag{4.8}$$

The time-dependent conduction current density, $\mathbf{j}(\mathbf{x};\mathbf{t})$, is then calculated by the following formula:

$$\mathbf{j}(\mathbf{x};\mathbf{t}) = \frac{q\hbar}{2m^*(\mathbf{x})i}\left(\frac{\partial}{\partial\mathbf{x}} - \frac{\partial}{\partial\mathbf{x}'}\right)\rho(\mathbf{x},\mathbf{x}';t)\Big|_{\mathbf{x}'=\mathbf{x}} \tag{4.9}$$

The collision term, $C_{col}\rho$, which introduces dissipation processes and leads to time-irreversibility of the system, is a non-trivial part of this theory. Several discussions have been reported on the collision term in the density matrix formalism. Firstly, an accurate quantum mechanical expression for electron–phonon interaction was reported by Levinson [8]. He derived a closed equation for the density matrix of electrons which are weakly interacting with equilibrium phonons. In his formula, however, the interaction term involves time-integrations (i.e. the past history of the system), for which numerical calculations are non-trivial. Secondly, a semi-classical model was proposed by Caldeira and Legget [9] relating to the theory of quantum Brownian motion:

$$C_{col}\rho(\mathbf{x},\mathbf{x}';t) = -\gamma(\mathbf{x} - \mathbf{x}')\left(\frac{\partial\rho(\mathbf{x},\mathbf{x}';t)}{\partial\mathbf{x}} - \frac{\partial\rho(\mathbf{x},\mathbf{x}';t)}{\partial\mathbf{x}'}\right)$$

$$- \frac{2m^*\gamma k_B T}{\hbar^2}(\mathbf{x} - \mathbf{x}')^2\rho(\mathbf{x},\mathbf{x}';t) \tag{4.10}$$

where γ is the coupling constant of electrons to a reservoir system. Although this expression is simple enough to be applied to numerical simulations, it has been shown that this model is only correct when the thermal energy, $k_B T$, is much larger than the coupling energy between the electrons and the reservoir. Thirdly, the simplest model for the collision term is a well-known relaxation-time approximation:

$$C_{col}\rho(\mathbf{x},\mathbf{x}';t) = \frac{\rho(\mathbf{x},\mathbf{x}';t) - \rho_{QE}(\mathbf{x},\mathbf{x}')}{\tau_s} \tag{4.11}$$

where ρ_{QE} is the quasi-equilibrium density matrix under an external bias, and τ_s is the phase-coherence breaking time. In the following calculations we choose the relaxation-time approximation because of its simplicity in numerical calculations. One difficulty with the use of this approximation for a real device is the fact that the distribution of electrons varies with position and external bias. Equation (4.11) has an effect on the position–space distribution of the electrons and, if the thermal equilibrium density matrix ρ_0 is used directly as ρ_{QE}, this results in breaking the current continuity. Obviously, an adequate model for ρ_{QE} is required which has the same electron distribution as $\rho(\mathbf{x},\mathbf{x}';t)$. The following approximation for ρ_{QE} has been proposed by Frensley [10] assuming Boltzmann statistics for the electrons:

$$\rho_{QE}(\mathbf{x},\mathbf{x}';t) = \sqrt{(\rho(\mathbf{x},\mathbf{x};t)\rho(\mathbf{x}',\mathbf{x}';t))}\exp\{-(\mathbf{x}-\mathbf{x}')^2/\lambda_T^2\} \tag{4.12}$$

This expression is easily found to be correct for thermally distributed free electrons, but its general justification might rely on the fact that eqn (4.12) reproduces the correct thermal equilibrium density matrix given by eqn (4.1) [10]. In the present work, we adopt eqn (4.12) for ρ_{QE} by adding a numerical correction for the difference between the Boltzmann and Fermi–Dirac distributions. The resultant expression gives us a quantitatively proper reproduction of off-diagonal elements of ρ_0 in thermal equilibrium.

For the numerical calculations, eqn (4.5) is discretised in $\mathbf{x}$ and $\mathbf{x}'$ using exactly the same spatial grid as the equilibrium density matrix (see Fig. 4.6) and is then deformed to algebraic equations. A Neumann-type boundary condition is implemented in order to conserve electron density at the device edge [10], [11]. For the time integral of eqn (4.6), an implicit scheme is used to obtain stable convergence in the numerical calculations since explicit schemes usually require extremely small time steps [10]. The coefficient matrix of the resulting simultaneous equations is basically sparse (although it is not symmetric and contains complex elements) and thus the conjugate gradient method for non-symmetric sparse matrices can be adopted. Values of 1.0–2.0 fs are used for the time step which are small enough to represent the transient response of the resonant tunnelling diode. The phase-coherence breaking time, τ_s, due to scattering is typically set to be 100 fs, which roughly equals the momentum relaxation time for n-GaAs with a donor concentration of 1×10^{18} cm^{-3} at room temperature. The density matrix, potential profile and current density are calculated using eqn (4.9) and these are monitored in order to check the convergence of the system to a steady state. The correctness of the thermal equilibrium density matrix as an initial condition is continuously checked by solving the density matrix equation under zero applied voltage. The maximum errors in the potential profile and current density under zero bias are less than 1.0 meV and 1.0 A cm^{-2} at a time of 1500 fs respectively. These are small enough to assure the propriety of evaluated values under an applied voltage.

4.2.2 Femtosecond electron dynamics in RTDs

Let us take a look at the transient behaviour of the density matrix under an applied bias by solving eqn (4.6) numerically. The calculated transient response of the density matrix for the resonant tunnelling diode is shown in Figs. 4.7 and 4.8. These diagrams show the evolution of the real (Fig. 4.7) and imaginary (Fig. 4.8) parts of the density matrix at time t after an external voltage of 120 mV, corresponding to the peak voltage of the present RTD, is applied at $t = 0$: (a) at $t = 50$ fs; (b) at $t = 100$ fs; (c) at $t = 200$ fs; (d) at $t = 600$ fs; and (e) at $t = 1000$ fs. The system reaches the steady state at a time of about 1000 fs. The calculated density matrix enables us to observe not only the time-dependent variation of the electron distribution but also that of the electron correlations in the system. The diagonal of the real parts shows the time-dependence of the electron build-up in the quantum well as well as that of the electron accumulation in the emitter region and depletion in the collector region. In the steady state, the peak concentration of the electrons in the quantum well amounts to 2.5×10^{17} cm^{-3}, which is about twice as large as that in thermal equilibrium. In Figs. 4.7 and 4.8, three major features should be noted in the off-diagonal elements of the density matrix. The large cross-diagonal structure in the imaginary parts (indicated by arrow A in Fig. 4.8(a)) represents short-range phase-

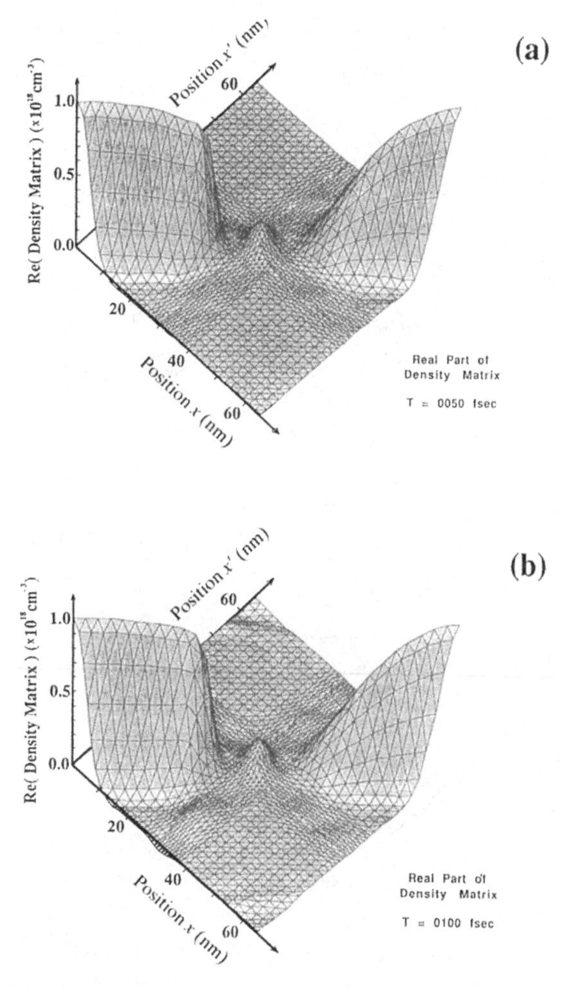

Figure 4.7 Real parts of the density matrix at various times under an applied bias of 120 mV: (a) at 50 fs; (b) at 100 fs; (c) at 200 fs; (d) at 600 fs; and (e) at 1000 fs. It can be seen that the density of the electrons accumulated in the quantum well increases with time. Electron depletion in the collector layer and accumulation in the emitter layer can also be seen [11]. (Figure 4.7(c)–(d) is on p. 102.)

correlations which cause plasma current oscillations. As shown later in this section, if the momentum–space representation is used instead of the position–space representation, this structure is equivalent to the centre of the Wigner distribution function oscillating along the momentum axis. In other words, electrons are in collective motion in the emitter and collector regions. This phenomenon can also be observed as an oscillation of the real electron density (the diagonal of Fig. 4.7).

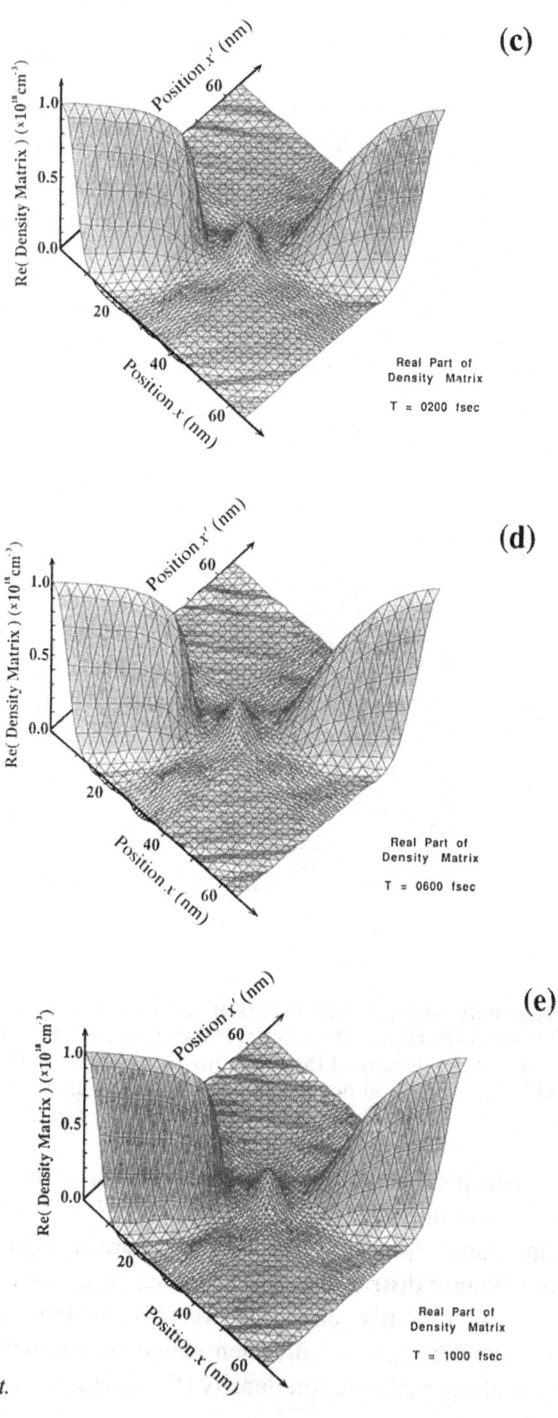

Figure 4.7 *cont.*

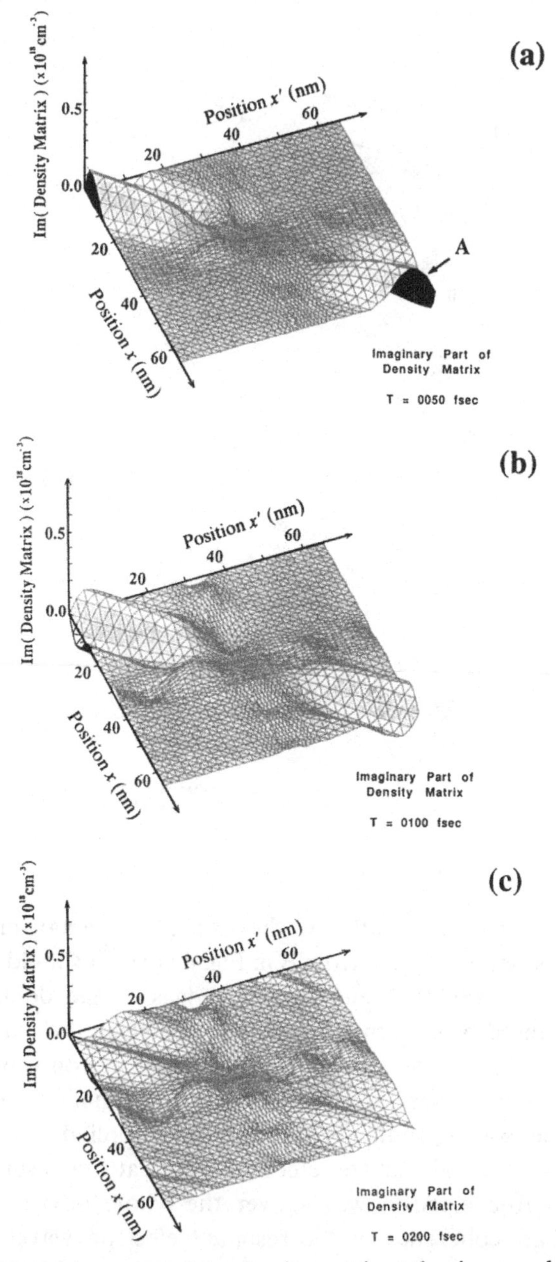

Figure 4.8 Imaginary parts of the density matrix at the times as shown in Fig. 4.7. The large cross-diagonal variation (indicated by A) represents a plasma oscillation. The remarkable oscillatory feature (indicated by B in (e)) measures the quantum interference between the resonant tunnelling electron and free electrons in the emitter and collector [11]. (Figure 4.8(d)–(e) is on p. 104.)

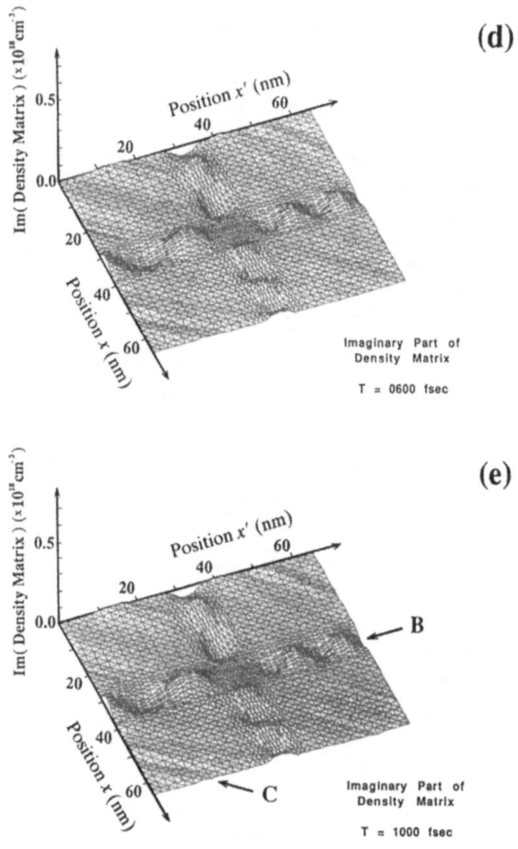

Figure 4.8 *cont.*

The most remarkable feature is the oscillatory behaviour of the imaginary parts (indicated by arrow B in Fig. 4.8(e)). It should be noted that this extends from the centre to the edges of the domain. This behaviour is caused by quantum interference between the quasi-bound electron waves in the resonant state and the free electron-waves in the emitter and collector layers, and demonstrates the dynamic correlations of the electron waves in the RTD under an applied voltage. The calculated results reveal that the electron waves at the resonant state correlate with free electron waves over the whole device. In other words, the phase-coherence of the resonant electron waves is maintained, although they are suffering from degradation of coherence due to scattering. The relationship between the amplitude of the oscillation and the phase-coherence breaking time is discussed later in this section.

Another oscillatory structure along the diagonal direction (indicated by arrow C in Fig. 4.8(e)) represents the correlations between the free electrons in the emitter and collector regions by normal tunnelling through the double barriers. Such correlations can be observed even in a single-barrier structure.

The corresponding transient current, calculated using eqn (4.9), is shown in Fig. 4.9, and its interpretation is given below. Strictly speaking,

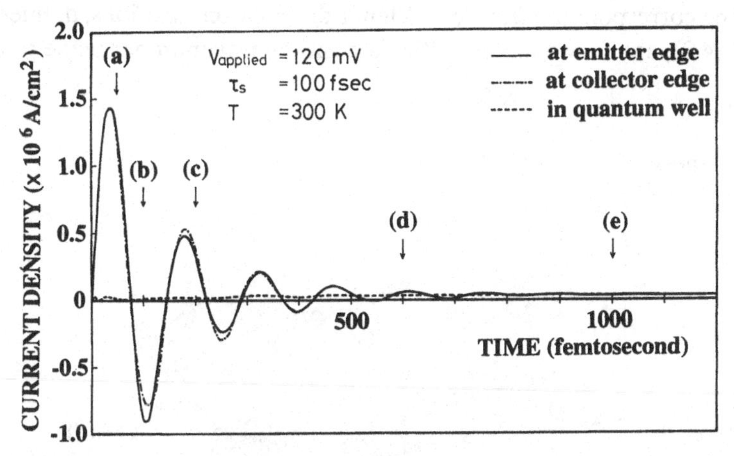

Figure 4.9 Time-dependence of current density calculated using eqn (4.9). The current densities at different positions in the device are monitored versus time: at the emitter edge (solid line); at the collector edge (broken line); and in the quantum well (dotted broken line). The arrows in the diagram ((a)–(e)) indicate the times at which the real and imaginary parts of the density matrix are drawn in Figs. 4.7 and 4.8.

the current density defined by eqn (4.9) is only correct in the steady state, since we do not include the time-dependent displacement current. Because of this simplification, the transient current density has a positional dependence in Fig. 4.9: the current densities at the emitter edge, the collector edge and in the quantum well are shown by a solid line, a broken line with a dot and a broken line respectively. Although it is possible to calculate the displacement current from the time-derivative of the self-consistent Hartree potential, we intentionally show only the time-dependence of the current caused by the change in electron density. The current density at the emitter edge (or the collector edge) can be regarded as the total current because the displacement current at the device edge is nearly zero. As described above, a large current oscillation is observed in Fig. 4.9 as a result of a plasma-type motion of electrons in the emitter and collector regions, as seen in the imaginary parts of the density matrix (Fig. 4.8(a)–(e)). The period of the current oscillation of about 150 fs is mainly determined by the density and effective mass of the electrons in the emitter and collector layers. The system reaches the steady state in about 100 fs and produces a steady current of 3.1×10^4 A cm^{-2}. It should be noted that the first peak current of the plasma oscillation is more than one order of magnitude larger than this steady current.

The corresponding transient Wigner distribution functions, defined by eqn (4.2), are shown in Fig. 4.10 in a position–wavenumber plane to give

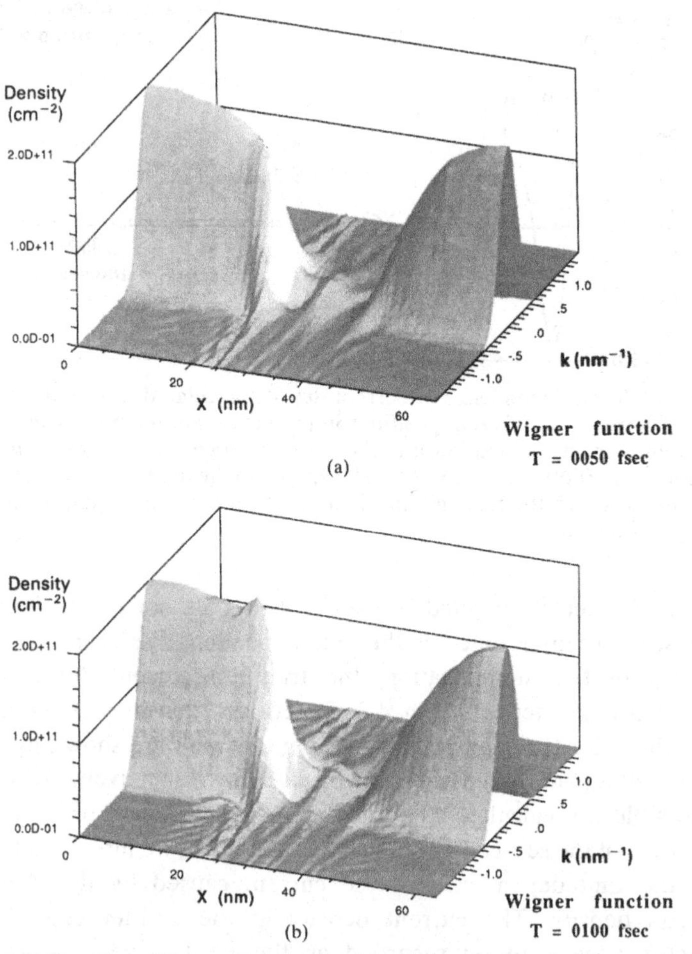

Figure 4.10 Transient Wigner distribution functions calculated at the same times as shown in Figs. 4.7 and 4.8 using eqn (4.2): (a) at 50 fs; (b) at 100 fs; (c) at 200 fs; (d) at 600 fs; and (e) at 1000 fs.

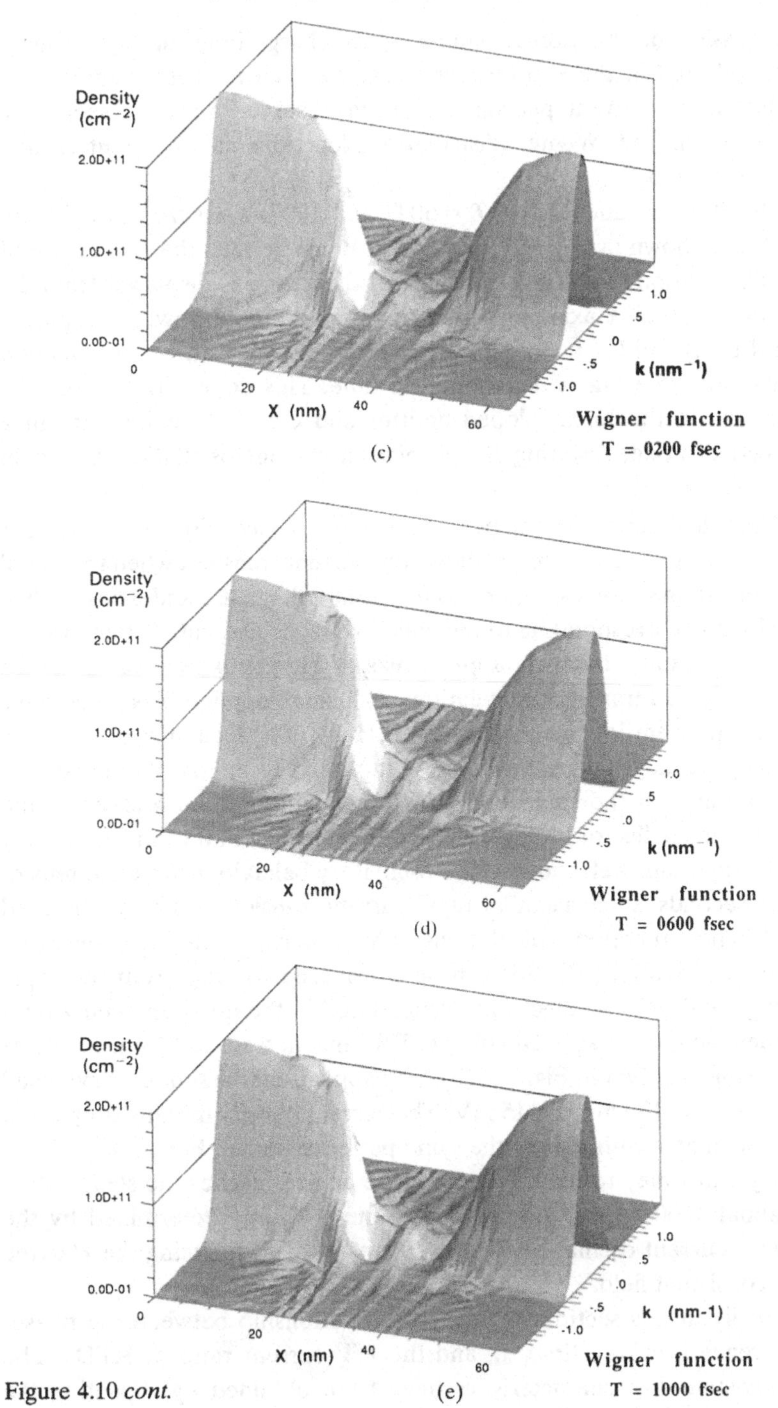

Figure 4.10 *cont.*

the classical distribution functions. Space charge build-up in the quantum well and in the emitter region can be seen in these diagrams. In addition, the current plasma oscillation observed above is seen as a variation in the Wigner distribution functions in the emitter and collector regions.

The Wigner functions at $T = 50$ fs and 100 fs seen from a different angle are shown in Fig. 4.11(a) and (b). It can be seen that the centre of the distribution function at the edge of the device is displaced from the origin of the k-axis representing the current flow. Comparing Fig. 4.11(a) and (b) it can be seen that the overall distribution function shifts along the k-axis depending on time. This demonstrates that the electrons in the highly doped emitter and collector regions are in a collective mode, resulting in the plasma current oscillation shown in Fig. 4.9.

The calculated switching behaviour of the device with spacer layers is shown in Fig. 4.12, where (a) shows the current transient when the initial bias of 115 mV, corresponding to the peak voltage, is suddenly switched to 170 mV, corresponding to the valley voltage. The initial state at $t = 0$ is a steady state under an applied bias of 115 mV which was obtained from repeated transient calculations with small applied bias increments as described earlier (see inset in Fig. 4.12(a)). The meanings of the three lines in the diagram are the same as those in Fig. 4.9. A plasma current oscillation with the same period as that in Fig. 4.9 is observed in the emitter and collector layers. The time-dependence of the current density in the quantum well shows some oscillatory behaviour for some tens of femtoseconds. Since a similar rapid current transient has been observed in Wigner function calculations which neglect the self-consistent Hartree potential [12], this can be considered to arise from the rapid change in density of electrons accumulated in the quantum well after a sudden increase in applied voltage. The current transient for the reverse switching is shown in Fig. 4.12(b). The applied bias is suddenly switched at $t = 0$ from 170 mV to 115 mV. The current density initially drops, and then oscillates with almost the same period as in (a). For both peak-to-valley and valley-to-peak switching, the device reaches the steady-state at about 1000 fs, and the switching time is mainly determined by the plasma current oscillation, which is a result of introducing the Hartree self-consistent field.

Finally in this section we study the relationship between the phase-coherence breaking time, τ_s, and the P/V current ratio of RTDs. The steady-state I–V characteristics have been obtained by repeating the

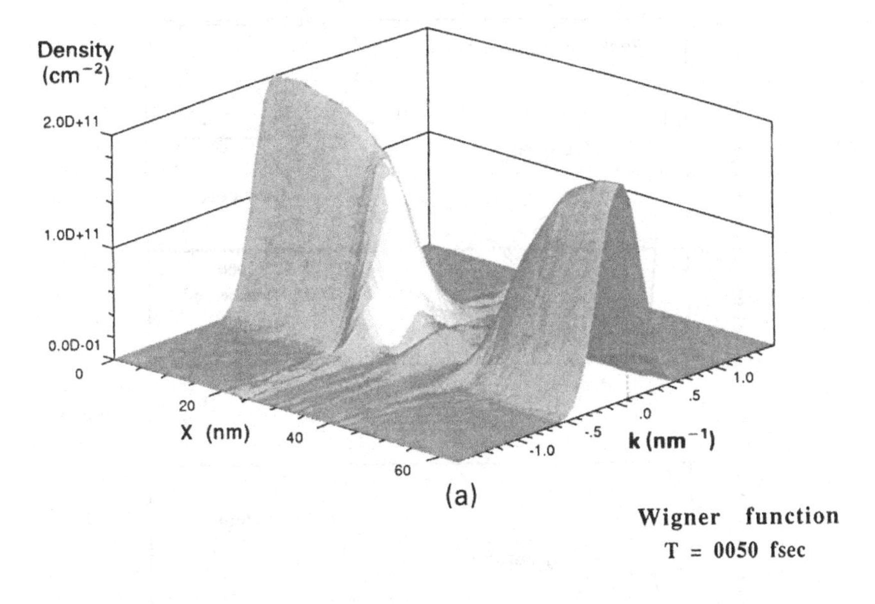

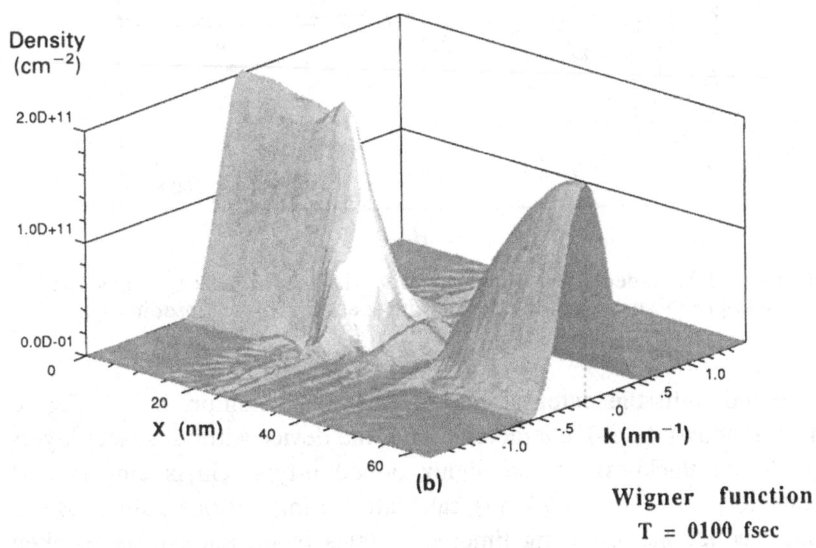

Figure 4.11 Wigner distribution functions at 50 fs (a) and 100 fs (b) seen from a different angle to those shown in Fig. 4.10.

transient calculations shown above. A small external voltage, typically 5–10 mV, is applied to the device in thermal equilibrium. The current value after a time of 1500 fs is stored as a steady-state current, and then the same step voltage is applied to this steady-state value. This process is

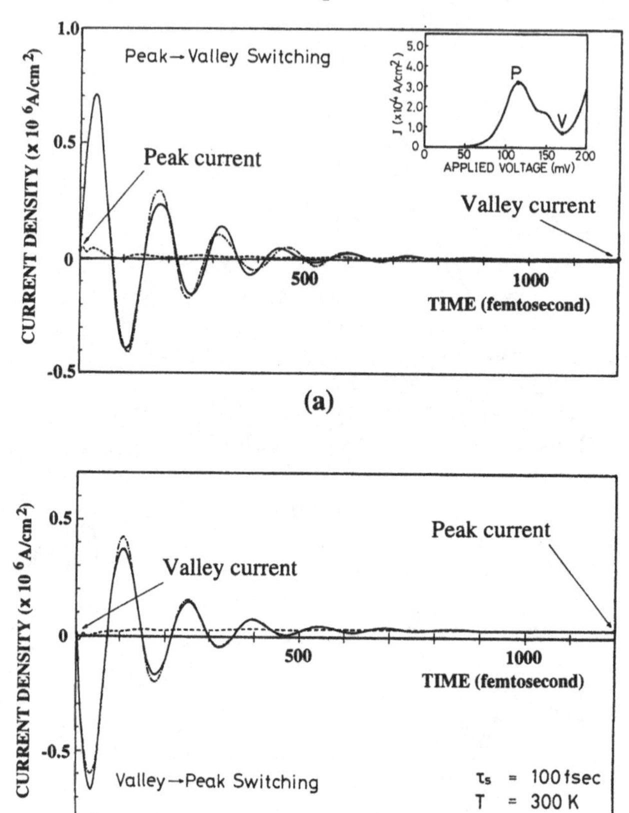

Figure 4.12 Time-dependence of current density for (a) peak-to-valley and (b) valley-to-peak switching of the device with space layers (see inset in (a)).

repeated until the external bias reaches its maximum value. Figure 4.13(a) shows the *I–V* characteristics of the device without spacer layers (with the thickness of the highly doped n-type GaAs emitter and collector layers set to 24 nm) calculated using various values of the phase-coherence breaking time: $\tau_s = 100$ fs (solid line); 80 fs (broken line); 60 fs (one-point broken line); and 40 fs (two-point broken line). As shown in this diagram, the peak current decreases, and the valley current rapidly increases as τ_s is reduced, resulting in a large degradation in the peak-to-valley current ratio. It should be noted that the *I–V* curve for a scattering time of 40 fs no longer has NDC region.

The strong τ_s-dependence of the *P*/*V* current ratio can be understood as follows: Fig. 4.13(b) shows the voltage dependence of the sheet

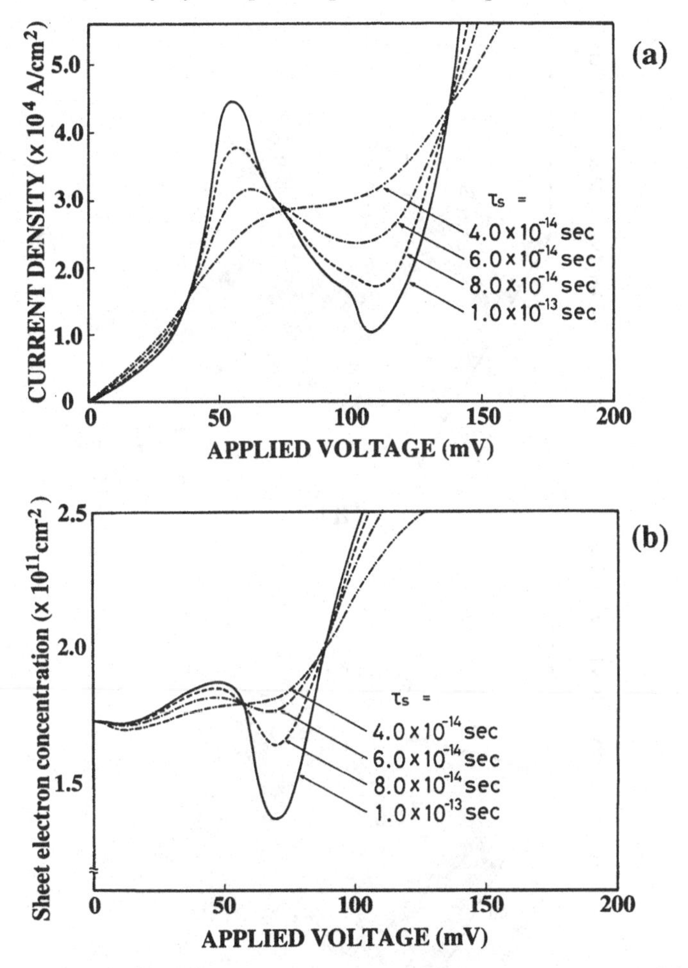

Figure 4.13 Applied voltage dependence of (a) current density and (b) sheet electron concentration in the quantum well calculated for the device without spacer layers for various values of τ_s.

concentration, n_S, of electrons accumulated in the quantum well calculated using various values of τ_s. It can be seen that the n_S–V curves have peak-to-valley structures corresponding to the NDC of the I–V characteristics, and that the peak-to-valley ratio also decreases with decreasing τ_s. A shorter τ_s corresponds to more frequent momentum relaxation of electrons in the quantum well. In other words, the phase-coherence of the electron waves at resonance degrades and the tunnelling mechanism changes from coherent tunnelling to sequential tunnelling. The increase in n_S for the valley state indicates that a number of electrons are still

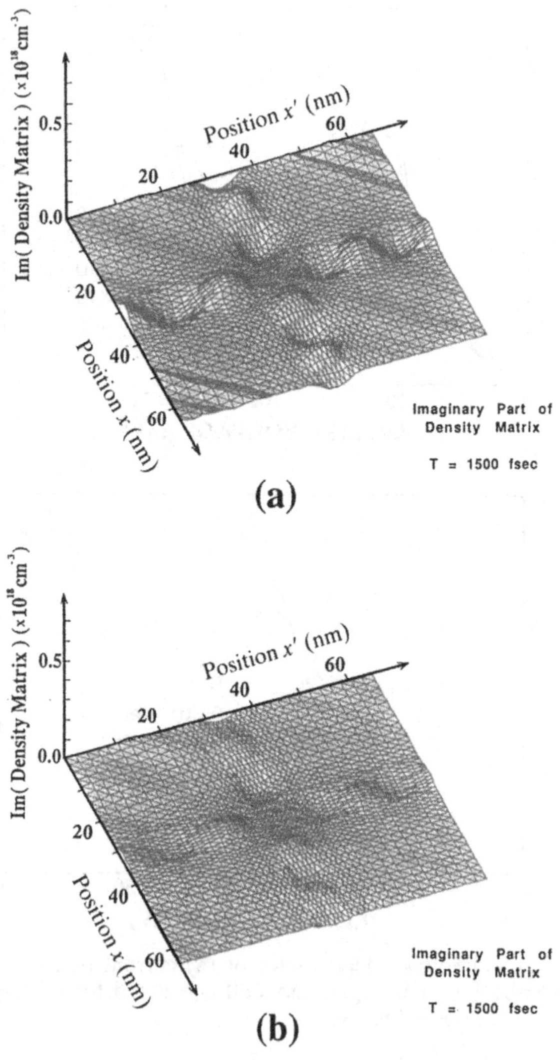

Figure 4.14 Imaginary part of the steady-state density matrix (at a time of 1500 fs) calculated with a τ_s of (a) 100 fs and (b) 40 fs. The reduced oscillatory behaviour of the off-diagonal elements in (b) signifies a degradation of the phase-coherence in the device due to a shorter τ_s.

present on the quasi-bound level even at off-resonance, resulting in a larger valley current.

Figure 4.14 shows the imaginary parts of the density matrix at the peak-current state calculated using a τ_s of (a) 100 fs and one of (b) 40 fs. As explained earlier, the phase-coherence of the electron waves at resonance can be observed as oscillatory structures in the off-diagonal elements. Comparing parts (a) and (b), it can clearly be seen that the amplitude of oscillation in Fig. 4.14(b) is much smaller than that in Fig. 4.14(a). The diagram illustrates directly the degradation of the electron-wave coherence due to scattering.

4.3 Space charge build-up in the quantum well, and intrinsic current bistability of RTDs*

4.3.1 Numerical analysis of intrinsic current bistability

As mentioned in Chapter 2, double-barrier RTDs frequently exhibit current hysteresis in the NDC regime, the origin of which has given rise to heated controversy over the past few years. One simple origin of such hysteresis is the external series resistance either in the contact regions or in the measurement circuit. The external series resistance has two effects on the characteristics of the NDC device: circuit oscillation and current hysteresis. The circuit oscillation can be removed by connecting a capacitor in parallel to the NDC device, as shown by Goldman *et al.* [15]. The current bistability caused by the series resistance is exceedingly common among NDC devices and is not a characteristic of RTDs at all; this is thus referred to as *extrinsic bistability* [13], [14] to distinguish it from another form of bistability, termed *intrinsic bistability*, which is more important from a physical point of view, although it has a much smaller effect than does external bistability. Intrinsic bistability is ascribed to the dynamic charge build-up in the quantum wells of RTDs.

Figure 4.15(a) and (b) [16] shows *I–V* characteristics near the first resonance for Materials 2 and 3 respectively. Both materials show clear hysteresis. The fact that the hysteresis is similar in both cases, even though the currents are different by a factor of 100, suggests that it cannot be attributed to the extrinsic effects of series resistance.

As discussed in Section 2.3, the sheet concentration of the electrons accumulated in the well, σ_w, can be expressed using the density of the resonant tunnelling current, J, and the electron dwell time, τ_d, of electrons in the well introduced in Section 2.2.1:

$$\sigma_w = \frac{J \cdot \tau_d}{e} \tag{4.13}$$

* Parts of this section and Section 4.4 are based on the PhD dissertation by Goodings [1].

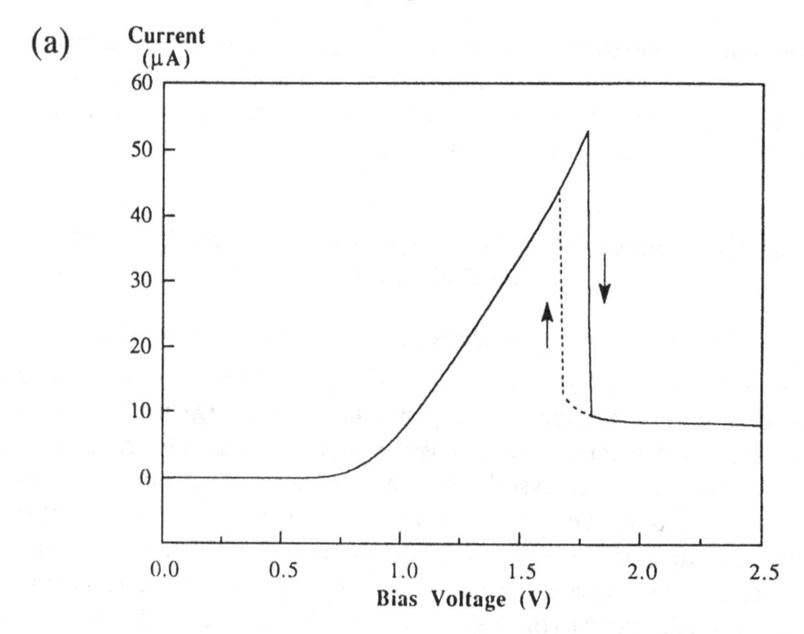

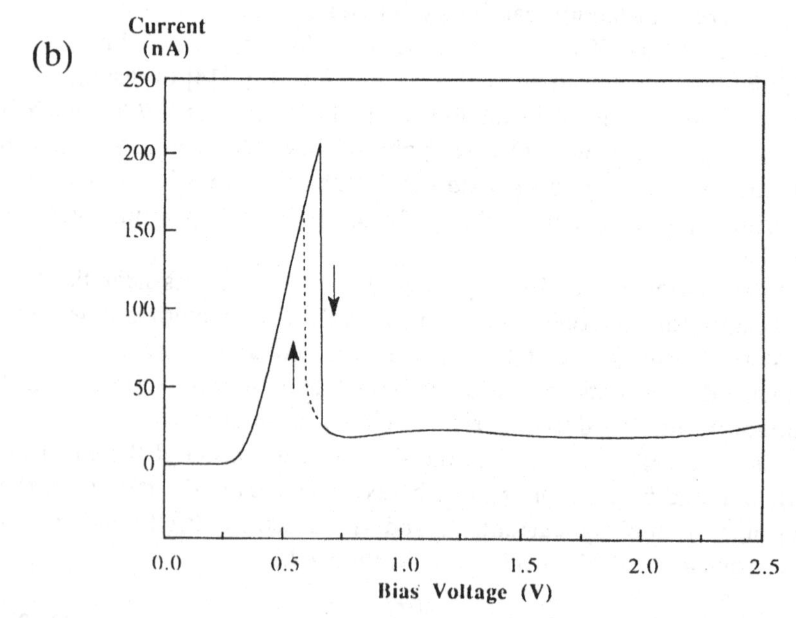

Figure 4.15 *I–V* characteristics for (a) Material 2 and (b) Material 3 at 4.2 K. The bias voltage was swept in both directions to show the hysteresis – the solid lines are for increasing voltage and the dashed line for decreasing voltage. Where the lines overlay, only the solid curve is shown. After Goodings *et al.* [16], with permission.

This means that σ_w has a feedback-like dependence on the tunnelling current. The electrostatic interaction between the electrons accumulated in the well gives rise to an energy shift of the quasi-bound state in the well, leading to a change in the tunnelling current. There is thus the possibility, at least qualitatively, of having two stable states of the device over a range of applied bias: a high current with a large accumulation and a low current with a small accumulation. Numerical calculations are obviously necessary for further discussions on this issue.

In parallel with experimental studies, several theoretical studies have been reported to demonstrate the possibility of space charge build-up. Analyses based on self-consistent calculations of the tunnelling current (see Section 2.2.2) by Berkowitz *et al.* [17] and Mains *et al.* [18], and a more analytical calculation, based on the rate equation for the electron occupancies in the well conducted by Sheard *et al.* [19], are pioneering studies which demonstrate theoretically the space charge build-up phenomenon inherent in RTDs. In these studies the electrons stored in the well were evaluated in two extreme situations, that is, global coherent tunnelling and incoherent sequential tunnelling limits. Recent numerical simulations [7], [11], [20] based on the quantum transport theory shown in Section 4.2 allow us to investigate the dynamics of tunnelling electrons in the more realistic intermediate regime. Figure 4.16 shows the *I–V* characteristics of the device calculated at room temperature using the density matrix theory described in Section 4.2.1. To observe any hysteresis in the *I–V* characteristics the external bias is first increased in the same manner as that for Fig. 4.13 and is then decreased by applying a small negative step voltage. In Fig. 4.16 the *I–V* characteristics are shown for both RTDs with and without the spacer layers. The peak voltage of the device with spacer layers becomes larger than that without spacer layers because of the additional voltage fall across them. The calculations in the NDC regions are found to be slightly unstable as compared with the positive differential resistance regions, and longer calculations are generally required to reach the steady state. Small intrinsic bistabilities in NDC regions are observed in both devices. In the case of the device with spacer layers it seems that, when bias is decreased, the *I–V* curve in the NDC region shifts towards a lower voltage. Figure 4.17 shows the energy-band diagrams corresponding to the two stable states of the device for an external bias of 0.15 V. When the bias is increased, electrons accumulate in the quantum well at this bias to a sheet concentration of 2.1×10^{11} cm^{-2}. However, in the case of decreasing bias, the electron concentration decreases to

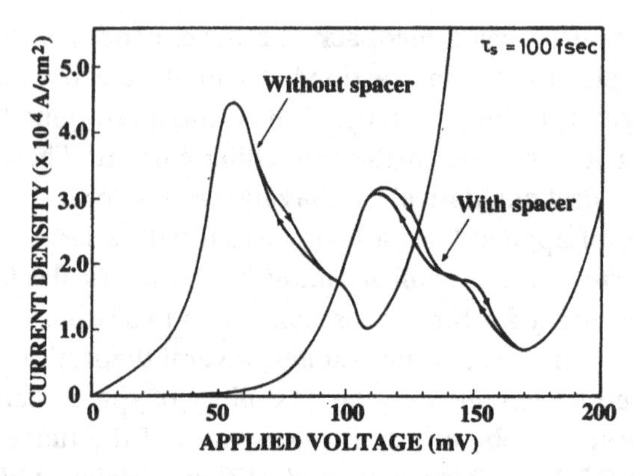

Figure 4.16 *I–V* characteristics calculated for RTDs with and without spacer layers. The applied bias is decreased after it reaches a maximum in order to see hysteresis. Bistability observed in the negative differential conductance region arises from dynamic electron redistribution in the quantum well.

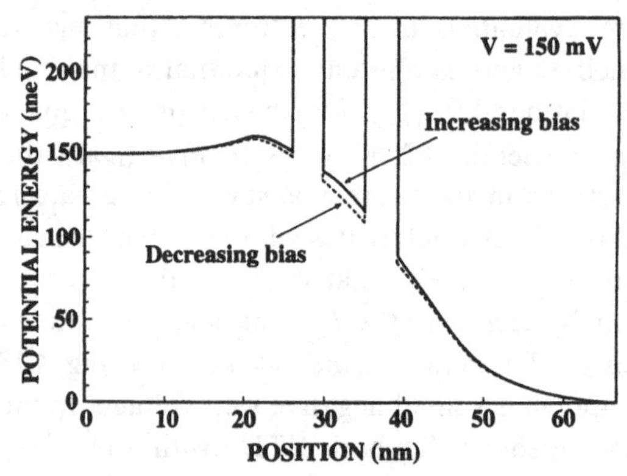

Figure 4.17 Self-consistent energy-band diagrams calculated at two stable states under an applied bias of 0.15 V. The upper curve corresponds to the larger current state, and the lower curve to the smaller current state.

1.5×10^{11} cm^{-2}. The difference in the space charge build-up leads to a different self-consistent Hartree field, and the bottom of the quantum well in the case of increasing bias is almost 10 meV higher than that in the case of decreasing bias.

Let us quickly take a look at the current shoulder in the NDC regions in Fig. 4.16. A similar structure has sometimes been observed experimentally in samples with large *P*/*V* current ratios, and several studies have reported [21], [22] on *extrinsic instability* due to the bias circuit.

From macroscopic circuit theory, it has been revealed that the *I–V* characteristics of an RTD with a bias circuit have a current shoulder whose shape depends on the LCR (inductance, capacitance, resistance) parameters of the external circuit [22]. Because the dynamics of electrons, not only in the quantum well but also in the emitter and collector are now simulated exactly, the present results might show the existence of *intrinsic instability* in the system. It is found that this current shoulder in the NDC region depends on the phase-coherence breaking time, τ_s, and generally vanishes with decreasing τ_s. This fact implies that the structure is produced by the electrons undergoing coherent tunnelling and the structure becomes small as the phase-coherence of the electron waves degrades with increasing scattering. However, the microscopic physical mechanism causing this structure is not clear at present, and further investigation is necessary.

Although the intrinsic bistability of symmetric double-barrier RTDs has been well attested theoretically, as shown here, its experimental observation is more disputable because of the extrinsic bistability. A practical experimental foundation of its existence was put forward by Alves *et al.* [23] and Leadbeater *et al.* [24] using an *asymmetric* double-barrier RTD. They investigated the *I–V* characteristics of an $Al_{0.4}Ga_{0.6}As$(8.3 nm)/GaAs(5.8 nm)/$Al_{0.4}GA_{0.6}As$(11.1 nm) double-barrier RTD in both bias directions and found that current bistability is observed only when the bias is applied so that the thicker barrier is on the collector side (i.e. the more positive side). We define an RTD biased in this way as being under reverse bias.

A schematic energy-band diagram is shown in Fig. 4.18(a) at the first resonance in the reverse bias direction. In these circumstances the electrons which tunnel through the thinner barrier are stored in the quantum well for a long period because of the following thicker barrier. This leads to a large space charge build-up. On the other hand, in the forward bias direction, the electrons accumulate much less in the quantum well since they tunnel through the thinner barrier quickly from the quantum well. As seen in Fig. 4.18(b) it was found that the *I–V* characteristics exhibit current bistability only in the reverse bias direction. This observation has confirmed the existence of intrinsic bistability caused by space charge build-up in the quantum well.

4.3.2 Magnetoconductance measurements of charge build-up

More direct evidence for space charge build-up in RTDs has been obtained by using the magnetotransport measurement technique: it has

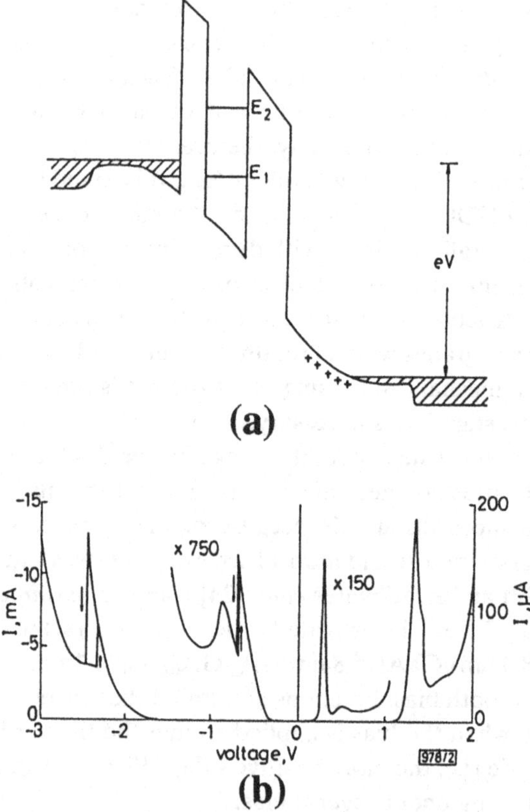

Figure 4.18 (a) Schematic energy-band diagram of an asymmetric double-barrier RTD. (b) *I–V* characteristics at 4 K of an asymmetric RTD measured at 4 K in both forward-(positive) and reversed-(negative) bias directions. After E. S. Alves *et al.* [23], with permission.

been demonstrated that the degree of electron accumulation can be determined by analysing the oscillations of current and capacitance seen in a magnetic field perpendicular to the barriers [25]–[29]. In this section this powerful technique is adopted for further investigation into space charge build-up and intrinsic bistability.

As we saw in Section 3.1.3, the lateral electronic states in both the accumulation region and the quantum well are given by replacing the lateral kinetic energy $\hbar^2 k_{//}^2/2m^*$ with Landau subband energy, and thus the total electron energy, E, is expressed in both regions as follows:

$$E = E_0 + (N + 1/2)\hbar\omega_c \qquad (4.14)$$

where E_0 is the lowest 2D state energy in the emitter or quantum well, N is the Landau-level quantum number, and ω_c is the cyclotron frequency ($=e\hbar B/m^*$). The condition for resonant tunnelling under a magnetic field is then simply expressed as follows:

$$E_0(\text{emitter}) + (N + 1/2)\hbar\omega_c = E_0(\text{quantum well}) + (N' + 1/2)\hbar\omega_c \quad (4.15)$$

The difference between N and N' represents the change in Landau level during the tunnelling. From eqn (4.14) the Nth Landau level aligns to the local Fermi energy, E_F^{local}, when the following condition is satisfied:

$$(1/B) = \left(N + \frac{1}{2}\right)\frac{e\hbar}{m^* E_F^{\text{local}}} \quad (4.16)$$

This means that the tunnel current, I, and conductance, G, versus B curves exhibit peaks whenever the Landau levels pass through the local Fermi level, leading to oscillations with a definite period in $1/B$. The interval between these peaks $\Delta(1/B)$ is thus related to the local Fermi energy E_F^{local} by the following equation:

$$(\Delta(1/B))^{-1} = \frac{m^* E_F^{\text{local}}}{e\hbar} \quad (4.17)$$

Therefore, the local Fermi energy can be extracted from the gradient of a graph of $(1/B)$ versus the Landau index number, N, which for this purpose is arbitrary. Furthermore, if we assume that all the Landau levels beneath the local Fermi level are full, that is, the electrons are completely thermalised, then the sheet concentration of the locally accumulated electrons σ_{local} can be deduced from the value of E_F^{local} obtained above:

$$E_F^{\text{local}} = \frac{\hbar^2 \pi \sigma_{\text{local}}}{m^*} \quad (4.18)$$

An example of magnetocurrent characteristics for Material 2 showing periodic structure is given in Fig. 4.19(a). The positions of the peaks (or troughs) have been extracted and plotted as $(1/B)$ versus an arbitrary index N (see inset). The data measured for Material 3 (see Fig. 4.19(b)) are in general clearer than those for Material 2, allowing the characteristics to be noted for a greater range of bias voltage. Thus for Material 3, results could be obtained for the charge build-up in the postresonance regime, while for Material 2 this was not the case. The symmetry of the peaks was seen to vary with the applied bias – an observation also made

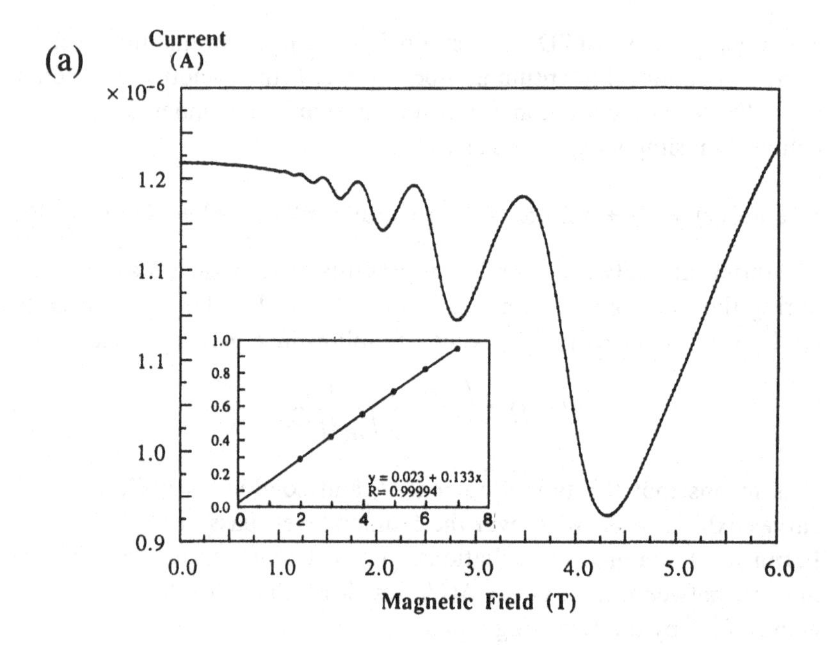

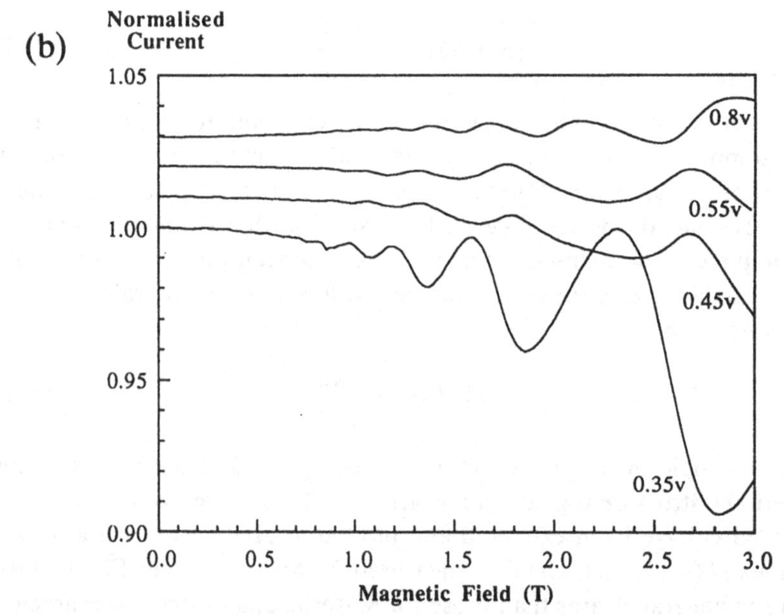

Figure 4.19 Magnetocurrent curves at 4.2 K (a) for Material 2, and (b) for Material 3. The inset of (a) shows a plot of $1/B$ versus the Landau index used to extract the Fermi energy and carrier density. Part (b) gives a set of normalised and offset curves for various bias voltages showing the asymmetry found at some biases. After Goodings [1], with permission.

by Thomas *et al.* [29]. Figure 4.19(b) shows a series of magnetocurrent curves that illustrate this effect for Material 3, normalised to their zero-field values and offset. The characteristics with greatest asymmetry appear in the range of bias between the threshold and peak resonance voltages (0.3–0.6 V; see Fig. 4.15(b)).

The bias dependence of the charge accumulation in the emitter has been successfully measured for Material 3 above the threshold voltage. The values obtained for E_F^e and σ_e in the emitter are shown in Fig. 4.20.

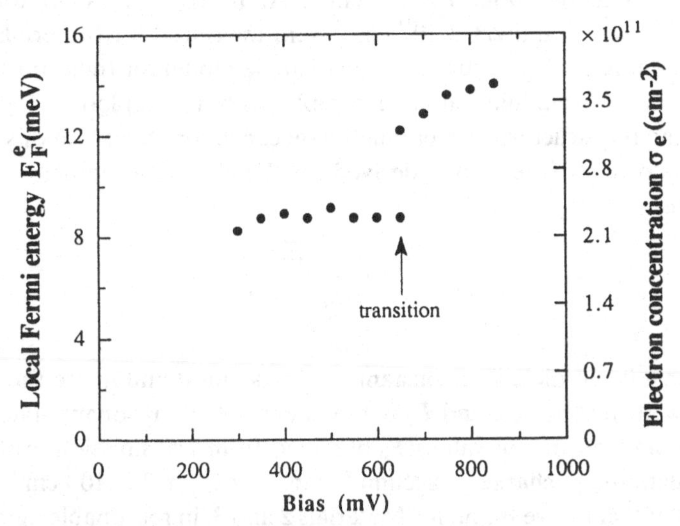

Figure 4.20 Energy spacing between the local Fermi level and the lowest 2D state in the accumulation layer and carrier density in the emitter region for Material 3. The results show a marked step in charge accumulation at the resonance peak (marked by the arrow). After Goodings *et al.* [16], with permission.

Although the charge build-up in the quantum well, σ_w, has not been measured directly in this experiment, it can be determined in the following way. The main process that occurs between the threshold voltage and the peak resonance voltage is the build-up of charge in the quantum well. This gives a large amount of electrostatic feedback, screening the emitter barrier and accumulation region from the increasing fields. Thus we would expect, and do see, that the Fermi energy E_F^e and the charge build-up in the emitter σ_e remain roughly constant in this range. As the bias is increased beyond the resonance peak, the resonant level in the quantum well falls below that in the emitter and thus the well suddenly becomes depopulated, leading to a step change in the bias across the emitter barrier and accumulation region. This gives a sudden change in E_F^e and σ_e, as shown in Fig. 4.20.

Electrostatic feedback provided by the voltage fall across the depletion region ensures that the charge lost from the quantum well is roughly balanced by the charge gained in the accumulation region, that is, $\sigma_w \approx \Delta(\sigma_e)$. In Fig. 4.20, we see that the charge in the accumulation region increases from 2.3×10^{11} cm^{-2} to 3.3×10^{11} cm^{-2} as the device switches, and thus charge accumulation in the quantum well, σ_w, is estimated to be about 1×10^{11} cm^{-2}. At resonance, values for σ_e of 3.7×10^{11} cm^{-2} and 2.3×10^{11} cm^{-2} were measured for Materials 2 and 3. Goldman *et al.* [30] suggest a very simple model for the estimation of the charge accumulation in a sample with no depletion region by treating the structure as a parallel-plate capacitor. Adapting this model, the following expression is derived for the charge accumulation in the emitter.

$$\sigma_e \approx \frac{\epsilon}{e} \frac{E_0}{d + w/2} \tag{4.19}$$

where ϵ is the dielectric constant of GaAs and d and w are the barrier and well thicknesses, and E_0 is the energy of the resonant state in the quantum well. Using values E_0 obtained from transmission probability calculations, charge accumulations of 6.2×10^{11} cm^{-2} and 3.5×10^{11} cm^{-2} are found for Materials 2 and 3, in reasonable agreement with, but about 50% higher than, the measured values.

An alternative technique for the measurement of charge build-up has been used by Leadbeater *et al.* [28]. The capacitance of an RTD will depend upon the charge contained since this affects the size of the depletion region, so that a periodic structure will also be observed in the magnetocapacitance data. This technique has the advantage that the charge can be measured even when there is no current flow, for example below threshold. However, the values of the capacitance are small, and for relatively small-area devices can easily be swamped by stray capacitances. Hence this method can only be used for large-area devices. By using this magnetocapacitance method, Leadbeater *et al.* were able to separate the charge build-up in the quantum well and the accumulation by taking the Fourier transform of the oscillatory structure. They measured the voltage dependence of the differential capacitance C (see Fig. 4.21(a)) of an asymmetric $Al_{0.4}Ga_{0.6}As$ (8.3 nm)/GaAs(5.8 nm)/ $Al_{0.4}Ga_{0.6}As$(11.1 nm) double-barrier RTD in the low-frequency regime

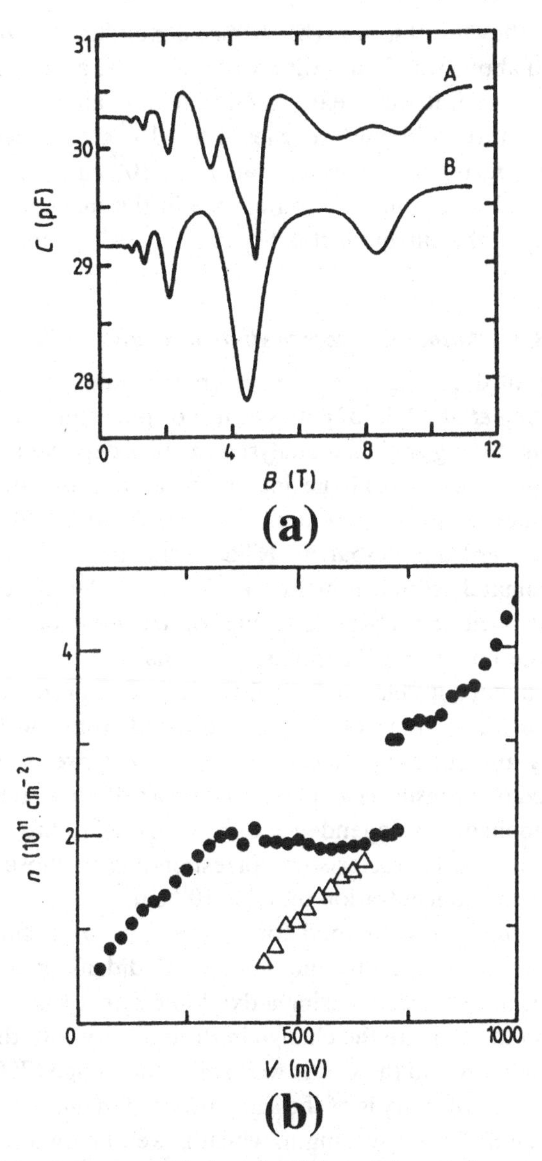

Figure 4.21 (a) Magneto-oscillations in differential capacitance C versus magnetic field B for applied voltage of 600 mV (curve A) and 300 mV (curve B). (b) Extracted areal density n versus voltage V for charge in the emitter (circles), and quantum well (triangles). After M. L. Leadbeater *et al.* [28], with permission.

from 10 kHz to 2 MHz. From the detailed Fourier analysis of the data, they found two series of oscillations which arise from the space charge build-up in the quantum well, σ_w, and in the emitter accumulation region, σ_e, as shown in Fig. 4.21(b). A bias dependence of σ_e similar to that discussed above was found with a step increase at the current peak voltage. In addition, it was clearly seen that the value of σ_w begins to increase from threshold and reaches $1.6 \times 10^{11}\ cm^{-2}$ near the peak voltage while σ_e remains constant at $1.9 \times 10^{11}\ cm^{-2}$ in this range. Leadbeater *et al.* also found the difference in the magneto-oscillations corresponding to the current bistability in the NDC regime.

4.3.3 Photoluminescence study of charge build-up

Another way of determining the space charge build-up, recently proposed by Young *et al.* [31], [32], makes use of photoluminescence (PL) measurements. Young *et al.* first analysed the bias dependence of the PL intensity, which is defined as the integrated area under the PL signal (see, for example, Fig. 4.23) from an $Al_{0.3}Ga_{0.7}As$(10 nm)/GaAs(5 nm)/ $Al_{0.3}Ga_{0.7}As$(10 nm) double-barrier RTD. Assuming that the PL intensity is approximately proportional to σ_w, namely holes are created only in the GaAs well, an absolute value of σ_w may be obtained by multiplying the integrated PL intensity, normalised to the signal from the structure under zero bias and the sheet electron density in the well at equilibrium, σ_w^0. The value of σ_w^0 was estimated from the background donor density in Young *et al.*'s analysis, but a more precise method has been proposed by Frensley *et al.* [33] based on a self-consistent screening model. The applied bias dependence of σ_w obtained in this way exhibits negative differential characteristics corresponding to those of the *I–V* curve with a peak concentration of $0.5 \times 10^{11}\ cm^{-2}$.

A crucial issue in this method for determining σ_w is how seriously minority holes contribute to the results. Vodjdani *et al.* [34], [35] performed PL measurements using a dye laser with a wavelength range of 700–800 nm to calculate the change in PL intensity with the energy of the exciting light around the GaAs well absorption edge (756 nm). They found that the PL intensity is of the same order of magnitude even when excited with an 800 nm wavelength, which is well below the absorption edge. This fact implies that minority holes are created in the anode region rather than in the quantum well and they then tunnel into the well resonantly. Therefore, Vodjdani *et al.* concluded that the tunnelling of holes is important and so it is difficult to obtain absolute values of σ_w from the PL intensity.

An alternative method of determining σ_w is to measure the bias dependence of the PL energy and line width. The PL linewidth has

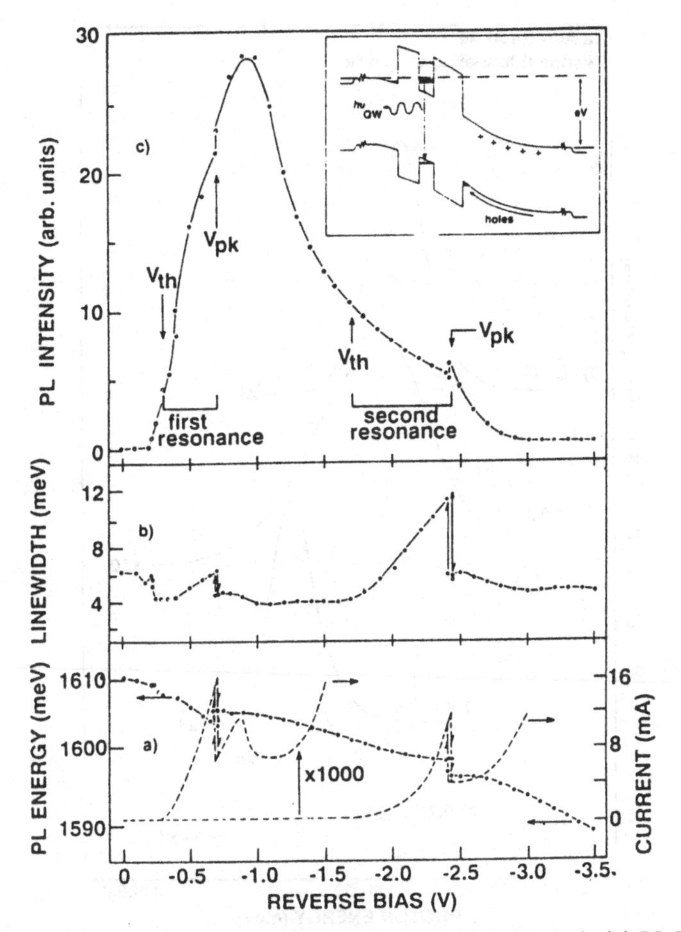

Figure 4.22 (a) Current (dashed) and PL peak position (circles); (b) PL FWHM; (c) integrated PL intensity as a function of reverse bias. The hysteresis loops are exhibited in the bias dependence of the current, PL peak energy and linewidth. The inset shows an energy-band diagram and quasi-Fermi level at the first resonance. After D. G. Hayes *et al.* [36], with permission.

already been shown to be useful in observing the space charge build-up in the experiments by Young *et al.* Figure 4.22 shows the experimental results reported by Hayes *et al.* [36] and Skolnick *et al.* [37] for an asymmetric $Al_{0.4}Ga_{0.6}As$(8.3 nm)/GaAs(5.8 nm)/$Al_{0.4}Ga_{0.6}As$(11.3 nm) double-barrier structure: the bias dependences of (a) current and PL energy, (b) PL linewidth and (c) PL intensity measured at 2 K. This device exhibits NDC regions due to the first and second resonances in the *I–V* characteristics at -0.7 and -2.44 V, with visible bistabilities

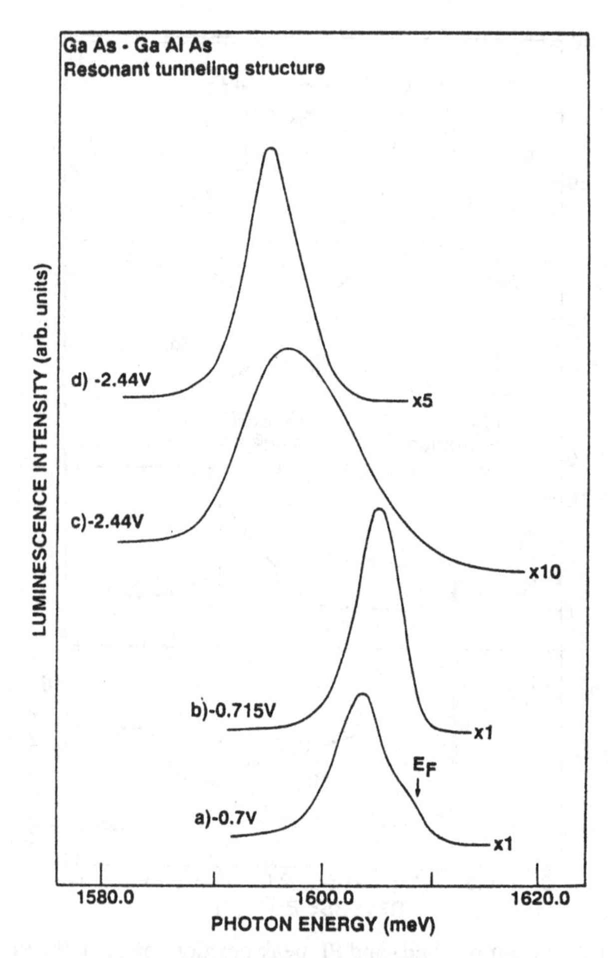

Figure 4.23 PL spectra in the high and low current states at the first resonance: (a) −0.7 V, −18.9 μA and (b) −0.715 V, −10.46 μA; and for the second resonance: (c) −2.44 V, −12.33 mA and (d) −2.44 V, −3.93 mA. The Fermi energy cut-off is indicated in (a). After D. G. Hayes *et al.* [36], with permission.

(Fig. 4.22(a)). Hysteresis can also be observed in the bias dependencies of the PL energy (Fig. 4.22(a)) and linewidth (Fig. 4.22(b)).

The original PL spectra [36], [37] are shown in Fig. 4.23 for bistable states at the first (Fig. 4.23(a), (b)) and second (Fig. 4.23(c), (d)) resonances. The PL peaks on-resonance (high-current state) shown in Fig. 4.23(a) and (c) are broader than those off-resonance and have tails towards higher bias. The difference in the PL linewidths arises from the free electron broadening of the distribution function [38] and therefore demonstrates the change in electron accumulation in the well. The process of deducing σ_w from the spectra is fairly complicated but, by convolving the PL line shape at the threshold of the first resonance where σ_w is thought to be negligibly small, Hayes *et al.* [36] have derived σ_w values of $(2.2 \pm 0.3) \times 10^{11}$ cm^{-2} and $(3.5 \pm 0.5) \times 10^{11}$ cm^{-2} for the first and second high-current (on-resonance) states. These are of the same order of magnitude as the values obtained theoretically.

4.4 Enhancement of current hysteresis under magnetic fields

In this section let us investigate the effect of a magnetic field applied perpendicular to the barriers on the hysteresis in the *I–V* characteristics. Figure 4.24(a) and (b) [16] shows the magnetic fields vs applied voltages where the current transitions are seen at 4.2 K in the *I–V* curves of Materials 2 and 3, respectively (see insets in Fig. 4.24(a)). The current hysteresis in the NDC regime exhibits a complex dependence on magnetic field. We see that for Material 2, the hysteresis is rapidly reduced with increasing field, with the positions of both the upper and lower transitions moving together until they meet at 4.0 T.

The insets in Fig. 4.24(a) show the actual form for the hysteresis curve, which differs at high and low fields. For Material 3, however, the field dependence of the hysteresis is completely different. For the range of 0–6 T studied, the lower transition remains roughly constant, while the upper transition moves to higher biases with some form of periodic structure being seen. Results similar to those for Material 2 have been observed and discussed by Eaves *et al.* [39], who have attributed the increase in hysteresis with applied field to rapid inter-Landau-level scattering in the quantum well which leads to an increase in the effective density of states. This in turn increases the capacity of the well to store charge without affecting the emitter barrier, so that a greater screening of the applied potential can occur and resonance will be obtained over a wider range of applied bias. Eaves *et al.* report that the increase in bias for the upper transition is not dramatic until all the electrons in the emitter region are contained in a single Landau level. According to the emitter areal charge density estimated from magnetotransport measurements (see Section 4.3.2), this will occur in Material 3 for fields above about 5.2 T, and thus such a dramatic increase will not be seen here. However, we can associate the periodic structure with the filling of the Landau levels in the emitter. Electrons tunnel from the emitter into

(a)

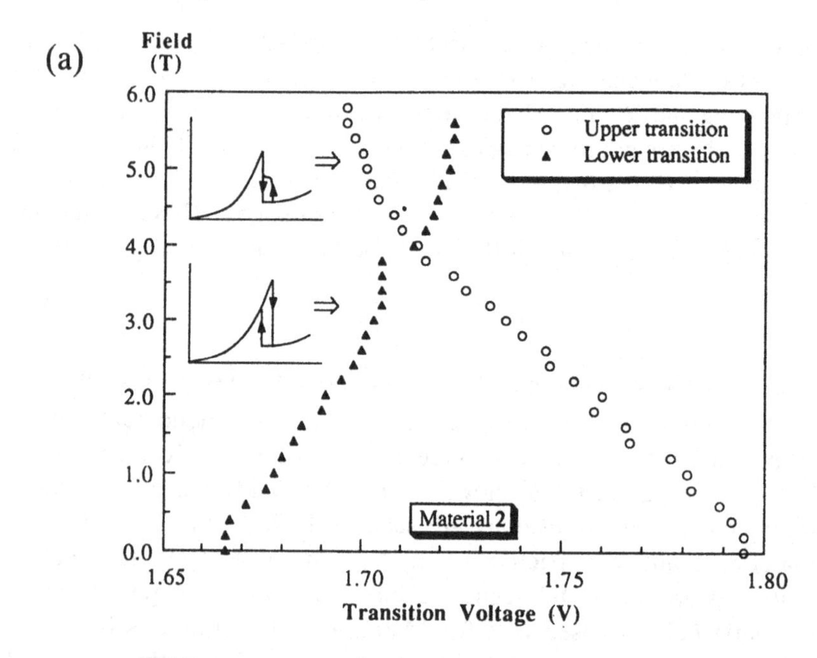

(b)

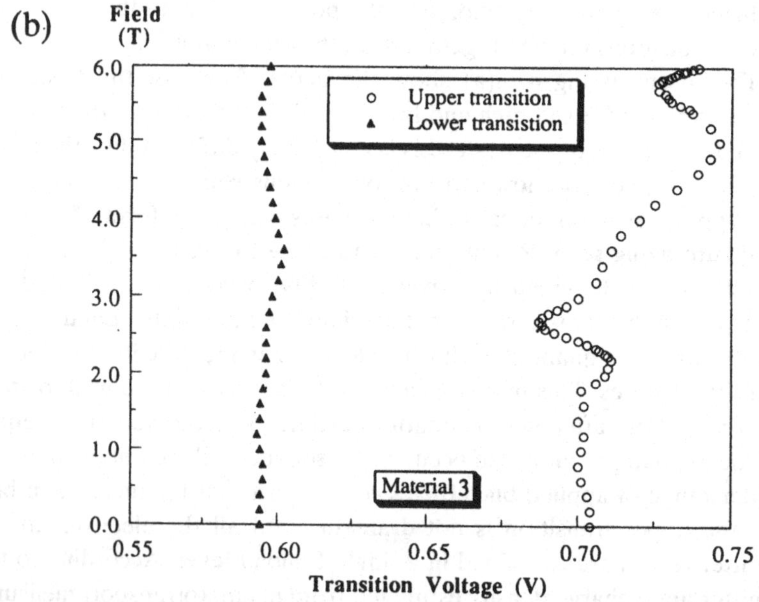

Figure 4.24 Positions of the current transitions as a function of magnetic field perpendicular to the barriers for (a) Material 2, and (b) Material 3. A reduction of hysteresis is seen with increasing field for Material 2, while an increase is seen for Material 3. The insets for (a) show the form of the *I–V* curve, which changes with applied field. After Goodings *et al.* [16], with permission.

empty states in the well from which the electrons relax. The number of empty states available to receive tunnelling electrons therefore depends upon the number of occupied (or partially occupied) Landau levels in the emitter and on the occupation of the levels in the well. As each successive Landau level in the emitter becomes depopulated, the number of empty states will be reduced and thus we might expect a minimum to occur in the amount of charge build-up and hence a minimum in the field-dependent hysteresis. Thus the minima in the hysteresis versus field curve of Fig. 4.24(b) should correspond to filled Landau levels in the emitter. The minima occur at 2.7 T and 5.7 T and, assuming an emitter electron density of 2.64×10^{11} cm^{-2} (10% higher than that measured, see Section 4.3.2), these values correspond to 0.86 and 2.02 filled Landau levels. We would not expect the agreement to be exact since the charge build-up in the accumulation region is bias- and field-dependent.

No such enhanced bistability is seen for Material 2. For the mechanism described above to hold, it is necessary for there to be sufficient time for the electrons in the quantum well to undergo inter-Landau-level transitions. Thus the electron lifetime in the well should be long compared with the lifetime of appropriate transitions, which could be due to acoustic phonon emission or electron–electron interactions. Ferreira and Bastard [40] give a lifetime of the order of 2×10^{-10} for intra-subband acoustic phonon transitions in similar systems. We can estimate the dwell time of electrons in the well, τ_d, from eqn (4.13). This gives values of 5×10^{-10}s and 1×10^{-7}s for Materials 2 and 3, respectively, and we see that for Material 3 the dwell time is indeed much longer than the intra-subband transition lifetime, as required. For Material 2, however, this is not the case, and could explain the difference in the observed characteristics.

4.5 References

[1] C. J. Goodings, Variable-area resonant tunnelling diodes using implanted gates, PhD thesis, Cambridge University, 1993.

[2] Y. G. Gobato, F. Chevoir, J. M. Berroir, P. Bois, Y. Guldner, J. Nagle, J. P. Vieren and B. Vinter, Magnetotunnelling analysis of the scattering processes in a double-barrier structure with a two-dimensional emitter, *Phys. Rev.*, **B43**, 4843, 1991.

[3] H. Zheng and F. Yang, Studies on tunneling characteristics on asymmetric GaAs/AlAs double-barrier structures, in *Resonant Tunneling in Semiconductors: Physics and applications*, edited by L. L. Chang, E. E. Mendez and C. Tejedor (Plenum, New York, 1990).

[4] F. Chevoir and B. Vinter, in *Resonant Tunneling in Semiconductors:*

Physics and applications, edited by L. L. Chang, E. E. Mendez and C. Tejedor (Plenum, New York, 1990).
[5] V. J. Goldman, D. C. Tsui and J. E. Cunningham, Observation of inelastic bistability in resonant tunneling structures, *Phys. Rev. Lett.*, **58**, 1256, 1987.
[6] D. Ter Haar, Theory and applications of the density matrix, *Reports on Progress in Physics*, **24**, 304, 1961.
[7] N. C. Kluksdahl, A. M. Kriman, D. K. Ferry and C. Ringhofer, Self-consistent study of the resonant-tunneling diode, *Phys. Rev.*, **B39**, 7720, 1989.
[8] I. B. Levinson, Translational invariance in uniform fields and the equation for the density matrix in the Wigner representation, *Sov. Phys.*, **JETP 30**, 362, 1970.
[9] A. O. Caldeira and A. J. Leggett, Path integral approach to quantum brownian motion, *Physica*, **121A**, 587–616, 1983.
[10] W. R. Frensley, Simulation of resonant-tunneling heterostructure devices, *J. Vac. Sci. Technol.*, **B3**, 1261, 1985.
[11] H. Mizuta and C. J. Goodings, Transient quantum transport simulation based on the statistical density matrix, *J. Phys.: Condens. Matter*, **3**, 3739, 1991.
[12] W. R. Frensley, Wigner-function model of a resonant-tunneling semiconductor device, *Phys. Rev.*, **B36**, 1570, 1987.
[13] T. C. L. G. Sollner, Comments on 'Observation of intrinsic bistability in resonant-tunneling structures', *Phys. Rev. Lett.*, **59**, 1622, 1987.
[14] T. J. Foster, M. L. Leadbeater, L. Eaves, M. Henini, O. H. Hughes, C. A. Payling, F. W. Sheard, P. E. Simmonds and G. A. Toombs, Current bistability in double-barrier resonant tunneling devices, *Phys. Rev.*, **B39**, 6205, 1989.
[15] V. J. Goldman, D. C. Tsui and J. E. Cunningham, Observation of inelastic bistability in resonant tunneling structures, *Phys. Rev. Lett.*, **58**, 1256, 1987.
[16] C. J. Goodings, H. Mizuta and J. R. A. Cleaver, Electrical studies of charge build-up and phonon-assisted tunnelling in double-barrier materials with thick spacer layers, *J. Appl. Phys.*, **75**, 2291, 1994.
[17] H. L. Berkowitz and R. A. Lux, Hysteresis predicted in *I–V* curve of heterojunction resonant tunneling diodes simulated by a self-consistent quantum method, *J. Vac. Sci. Technol.*, **B5**, 967, 1987.
[18] R. K. Mains, J. P. Sun and G. I. Haddad, Observation of intrinsic bistability in resonant tunneling diode modeling, *Appl. Phys. Lett.*, **55**, 371, 1988.
[19] F. W. Sheard and G. A. Toombs, Space-charge buildup and bistability in resonant-tunneling double-barrier heterostructures, *Appl. Phys. Lett.*, **52**, 1228, 1988.
[20] K. L. Jensen and F. A. Bout, Numerical simulation of intrinsic bistability and high-frequency current oscillations in resonant tunnelling structures, *Phys. Rev. Lett.*, **66**, 1078, 1991.
[21] Jeff. F. Young, B. M. Wood, H. C. Liu, M. Buchanan, D. Landheer, A. J. SpringThorpe and P. Mandeville, Effect of circuit oscillations of the dc current–voltage characteristics of double barrier resonant tunneling structures, *Appl. Phys. Lett.*, **52**, 1398, 1988.
[22] C. Y. Belhadj, K. P. Martin, S. Ben Amor, J. J. L. Rascol, R. C. Potter, H. Hier and E. Hempfling, Bias circuit effects on the current–voltage

characteristic of double-barrier tunneling structures: Experimental and theoretical results, *Appl. Phys. Lett.*, **57**, 58, 1990.
[23] E. S. Alves, L. Eaves, M. Henini, O. H. Hughes, M. L. Leadbeater, F. W. Sheard, G. A. Toombs, G. Hill and M. A. Pate, Observation of intrinsic bistability in resonant tunnelling devices, *Electron. Lett.*, **24**, 1190, 1988.
[24] M. L. Leadbeater, E. S. Alves, L. Eaves, M. Henini, O. H. Hughes, F. W. Sheard and G. A. Toombs, Charge build-up and intrinsic bistability in an asymmetric resonant-tunnelling structure, *Semicond. Sci. Technol.*, **3**, 1060, 1988.
[25] V. J. Goldman, D. C. Tsui and J. E. Cunningham, Resonant tunneling in magnetic fields: Evidence for space-charge buildup, *Phys. Rev.*, **B35**, 9387, 1987.
[26] L. Eaves, G. A. Toombs, F. W. Sheard, C. A. Payling, M. L. Leadbeater, E. S. Alves, T. J. Foster, P. E. Simmonds, M. Henini, O. G. Hughes, J. C. Portal, G. Hill and M. A. Pate, Sequential tunneling due to intersubband scattering in double-barrier resonant tunneling devices, *Appl. Phys. Lett.*, **52**, 212, 1988.
[27] C. A. Payling, E. S. Alves, L. Eaves, T. J. Foster, M. Henini, O. H. Hughes, P. E. Simmonds, F. W. Sheard, G. A. Toombs and J. C. Portal, Evidence for sequential tunnelling and charge build-up in double barrier resonant tunnelling devices, *Surface Science*, **196**, 404, 1988.
[28] M. L. Leadbeater, E. S. Alves, F. W. Sheard, L. Eaves, M. Henini, O. H. Hughes and G. A. Toombs, Observation of space-charge build-up and thermalization in an asymmetric double-barrier resonant tunnelling structure, *J. Phys: Condense. Matter*, **1**, 10605, 1989.
[29] D. Thomas, Magneto-tunneling studies of charge build-up in double barrier diodes, *Superlattices and Microstructures*, **5**, 219, 1989.
[30] V. J. Goldman, Bo Su and J. E. Cunningham, Resonant tunneling from an accumulation layer: New Spectroscopy of 2D electron systems, in *Resonant Tunneling in Semiconductors: Physics and applications*, edited by L. L. Chang, E. E. Mendez and C. Tejedor (Plenum, New York, 1990).
[31] J. F. Young, B. M. Wood, G. C. Aers, R. L. S. Devine, H. C. Liu, D. Landheer, M. Buchanan, A. J. SpringThorpe and P. Mandeville, Determination of charge accumulation and its characteristic time in double-barrier resonant tunneling structures using steady-state photoluminescence, *Phys. Rev. Lett.*, **60**, 2085, 1988.
[32] J. F. Young, B. M. Wood, G. C. Aers, R. L. S. Devine, H. C. Liu, D. Landheer, M. Buchanan, A. J. SpringThorpe and P. Mandeville, Photoluminescence characterization of vertical transport in double barrier resonant tunneling structures, *Superlattices and Microstructures*, **5**, 411, 1989.
[33] W. R. Frensley, M. A. Reed and J. H. Luscombe, Photoluminescent determination of charge accumulation in resonant tunneling structures, *Phys. Rev. Lett.*, **62**, 1207, 1989.
[34] N. Vodjdani, F. Chevoir, D. Thomas, D. Cote, P. Bois, E. Costard and S. Delaitre, Photoluminescence and space-charge distribution in a double-barrier diode under operation, *Appl. Phys. Lett.*, **55**, 1528, 1989.
[35] N. Vodjdani, D. Cote, D. Thomas, B. Sermage, P. Bois, E. Costard and D. Nagle, Electrical and optical evidence of resonant tunneling of holes in an n^+in^+ double-barrier diode structure under illumination, *Appl. Phys. Lett.*, **56**, 33, 1990.

[36] D. G. Hayes, M. S. Skolnick, P. E. Simmonds, L. Eaves, D. P. Halliday, M. L. Leadbeater, M. Henini, O. H. Hughes, G. Hill and M. A. Pate, Optical investigation of charge accumulation and bistability in an asymmetric double barrier resonant tunneling heterostructure, *Surface science*, **228**, 373, 1990.

[37] M. S. Skolnick, D. G. Hayes, P. E. Simmonds, A. W. Smith, H. J. Hutchinson, C. R. Whitehouse, L. Eaves, M. Henini, O. H. Hughes, M. L. Leadbeater and D. P. Halliday, Electronic processes in double-barrier resonant tunneling structures studied by photoluminescence spectroscopy in zero and finite magnetic fields, *Phys. Rev.*, **B41**, 10754, 1990.

[38] M. S. Skolnick, K. J. Nash, M. K. Saker, S. J. Bass, P. A. Claxton and J. S. Roberts, Free-carrier effects on luminescence linewidth in quantum wells, *Appl. Phys. Lett.*, **50**, 1885, 1987.

[39] L. Eaves, T. J. Foster, M. L. Leadbeater and D. K. Maude, Charge buildup, intrinsic bistability and energy relaxation in resonant tunneling structures: High pressure and magnetic field studies, in *Resonant Tunneling in Semiconductors: Physics and applications*, edited by L. L. Chang, E. E. Mendez and C. Tejedor (Plenum, New York, 1990).

[40] R. Ferreira and G. Bastard, Evaluation of some scattering times for electrons in unbiased and biased single- and multiple-quantum-well structures, *Phys. Rev.*, **B40**, 1074, 1989.

5

High-speed and functional applications of resonant tunnelling diodes

In contrast to the preceding chapters, which concentrated mainly on the physics of RTDs, this chapter reviews some applications of RTDs and related three-terminal devices. As briefly described in the first chapter, RTDs have two distinct features over other semiconductor devices from an applications point of view: namely, their potential for very-high-speed operation and their negative differential conductance. The former feature arises from the very small size of the resonant tunnelling structure along the direction of carrier transport; because of the short distance through which carriers must travel, RTDs can be designed to have very high cut-off frequencies. As a result, oscillation in submillimetre wave frequencies has been reported [1], [2]. Besides this high-speed potential, the negative differential conductance makes it possible to operate RTDs as so-called functional devices, which enables circuits to be designed on different principles than conventional devices. For example, signal processing circuits with a significantly reduced number of devices and multiple-valued memory cells using RTDs have been proposed and demonstrated. These functional applications are highly promising since RTDs, with their simple structure and small size, can be easily integrated with conventional devices such as field effect transistors (FETs) and bipolar transistors.

In Section 5.1, high-speed applications, including high-frequency signal generation and high-speed switching, are discussed. Functional applications, such as a one-transistor static random access memory (SRAM) and a multi-valued memory circuit, are described in Section 5.2.

5.1 High-speed applications of RTDs

This section deals with applications of RTDs in the millimetre and submillimetre frequency ranges. Because of their extremely non-linear *I–V* characteristics and the very short dwell time of electrons in the structure, RTDs have various potential applications at high frequencies. Reports on high-frequency signal generation are first reviewed in Section 5.1.1. Then high-speed switching, which is another promising application of RTDs, and its application to triggering circuits, are studied in Section 5.1.2.

5.1.1 High-frequency signal generation with RTDs

As described in preceding chapters, RTDs exhibit significantly non-linear *I–V* characteristics, namely, negative differential conductance (NDC), resulting from the quantum mechanical nature of electrons in double-barrier heterostructures. It was also shown in Chapter 4, by theoretical analysis, that the time constant of an RTD for switching from its current peak to valley, or from valley to peak, can be less than 1 ps. Furthermore, the capacitance of complete RTDs can be rather arbitrarily designed by choosing both the structural parameters of the double barrier and of the surrounding doping profile, thus changing the depletion layer thicknesses. This is in contrast to Esaki tunnel diodes in which the diode capacitance is very high since, in order to create NDC, the doping level must be sufficiently high for the semiconductor to degenerate. From an applications standpoint, these RTD properties are favourable in generating high-frequency signals. The negative differential conductance can provide the gain necessary for oscillation, and the smaller intrinsic delay and diode capacitance enables the diode to oscillate at higher frequencies. There have been reports of oscillators at frequencies never reached by other semiconductor devices [1], [2].

There have been thorough reviews on this subject by experimenters who have led the research in this field [3], [4]. Therefore, this section aims to provide only a general understanding of the operation of RTDs in high-speed circuits and its present status.

In order to understand the circuit operation of RTDs, let us consider an equivalent circuit of an RTD proposed by Brown *et al.* [5] and shown in Fig. 5.1. This small-signal-equivalent circuit of an RTD includes the following elements: firstly, a series resistance R_s arising from ohmic contacts, the resistivity of the emitter and collector regions, and spreading resistance; secondly, the parallel capacitance, C_D, resulting

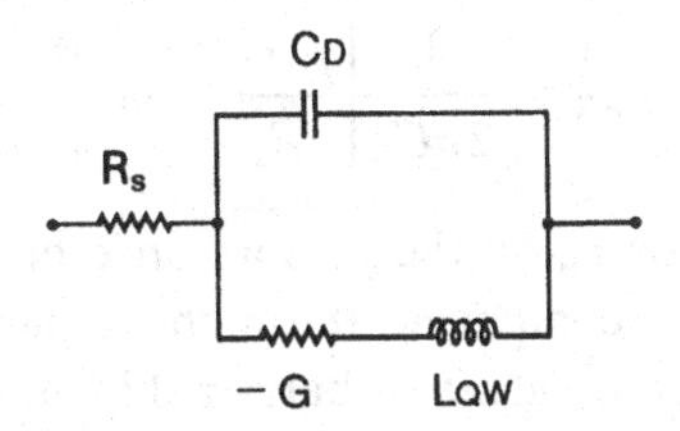

Figure 5.1 Small-signal equivalent circuit of a resonant tunnelling diode after Brown *et al.* [5].

from the charging and the discharging of electrons at the wiring and semiconductor depletion regions; and, finally, the differential conductance, $-G$, of the diode. The conductance in Fig. 5.1 has a negative sign which expresses the NDC of the RTD. In addition to the conventional equivalent circuit of a tunnel diode, an inductance element, L_{QW} is introduced. This inductance corresponds to the delay of the diode current with respect to the voltage [6], [7], arising from the time necessary to build up the charge in the quantum well in order to change the diode current. In general, L_{QW} is chosen so that the product of L_{QW} and G is equal to the electron dwell time, τ_d (see Section 2.3).

The maximum frequency of oscillation of the diode, f_{MAX}, is determined as follows. The impedance, $Z(f)$, of a diode as a function of frequency, f, can be expressed in terms of its real part, $R(f)$, and imaginary part, $X(f)$, in the following general formula:

$$Z(f) = R(f) + jX(f) \tag{5.1}$$

where j is used to express the imaginary number. When a current, I, flows into the diode, the electrical power consumed in the diode is given by $R(f) \cdot I^2$. Therefore, if $R(f)$ is negative, the diode can supply electrical power to the external circuit at the frequency, f, thus exhibiting gain. Consequently, f_{MAX} is defined as the frequency at which $R(f)$ becomes zero. For the equivalent circuit given in Fig. 5.1, f_{MAX} is given by [3]:

$$f_{MAX} = \frac{1}{2\pi} \cdot \left(\frac{1}{2{L_{QW}}^2 C_D} \right)^{\frac{1}{2}}$$

$$\left\{ 2L_{QW} - \frac{C_D}{G^2} + \left[\left(\frac{C_D}{G^2} - 2L_{QW} \right)^2 - \frac{4L_{QW}(1 + R_s G)}{R_s G} \right]^{\frac{1}{2}} \right\}^{\frac{1}{2}} \tag{5.2}$$

If L_{QW} is negligibly small, f_{MAX} is simply expressed in terms of the diode capacitance and the series resistance as follows:

$$f_{MAX} = \frac{1}{2\pi C_D}\left[\frac{-G}{R_s} - G^2\right]^{\frac{1}{2}} \tag{5.3}$$

The effect of L_{QW}, or τ_d, on the performance of RTD oscillators has been clearly shown by comparing the high-frequency performance of two diodes with different values of barrier thickness [3]–[5]. In Fig. 5.2

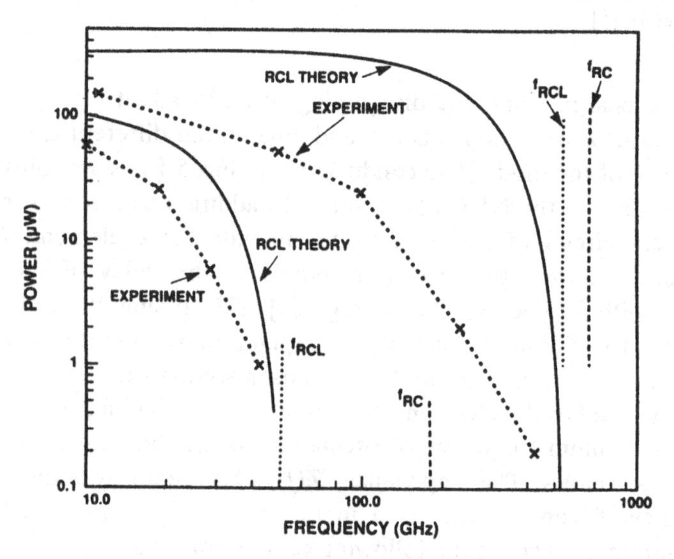

Figure 5.2 Output power as a function of frequency for two different double-barrier structures and their theoretical limits. The quantum well inductance is included in f_{RCL} but not in f_{RC} [3]. Reprinted with permission of MIT Lincoln Laboratory, Lexington, Massachusetts.

[3], the experimental output power of RTD oscillators is shown as a function of frequency along with their theoretical limits. Also shown are the f_{MAX} of the diodes derived from eqns (5.2) and (5.3) and denoted by f_{RCL} and f_{RC} respectively. It is clear from the diagram that theoretical results obtained from the equivalent circuit with L_{QW} agree with the experimental results better than those without L_{QW}. It is also clear that a reduction in L_{QW} is effective to increase f_{MAX}, enabling oscillation at very high frequencies up to 420 GHz [1].

Equations (5.2) and (5.3) also indicate that the high-frequency performance of RTDs can alternatively be improved by increasing G or decreasing R_s or C_D. The increase in G is strongly related to the reduction of L_{QW}. To obtain a large value of G, it is necessary to achieve a high peak current density, J_p, and a low valley current density, J_v. The value of J_p ranges over several orders of magnitude depending on the structural parameters of the double-barrier heterostructure, whereas the peak-to-valley current ratio, J_p/J_v, is around 10 for RTDs with high J_p. Thus to pursue high J_p is a promising way of realising large G. High J_p can be expected for RTDs consisting of double-barrier structures with a high transmission probability and thus a shorter electron dwell time. Therefore, for high-speed operation of RTD oscillators, high J_p is needed both to reduce the quantum well inductance and to increase G. There have been reports of two RTDs with a very high J_p: one has 450 kA cm^{-2} with J_p/J_v of 4 in an InGaAs/AlAs RTD [8] and the other has J_p of 370 kA cm^{-2} and J_p/J_v of 3.5 in an InAs/AlSb RTD [9]. In both cases, not only J_p but also the available current density, $J_p - J_v$, are very large at 345 kA cm^{-2} and 260 kA cm^{-2} respectively. These RTDs have an averaged specific NDC of more than 650 kS cm^{-2} with a NDC voltage of about 0.4 volts. Assuming a specific C_D of 100 nF cm^{-2}, which corresponds to a depletion layer thickness of a little more than 100 nm, a time constant, C_D/G, of 0.15 ps is obtained. Thus, for these RTDs, an oscillation frequency of 1 THz can be expected, provided that other parasitics are negligible.

In order to reduce C_D, the depletion layer must be made thick by employing a lower doping density on the collector side of the diode; however, as the total device thickness increases, the transit time of electrons across the whole structure becomes significant. Study of the influence of transit-time effects on RTD oscillators [10] has shown that the specific impedance Z_{tt} of the transit region at angular frequency ω can be expressed by the following:

$$Z_{tt} = \frac{W}{j\omega\epsilon}\left[1 - \frac{\sigma}{\sigma + j\omega\epsilon} \cdot \frac{1 - \exp(-j\theta_d)}{j\theta_d}\right] \tag{5.4}$$

where ϵ is the dielectric constant, W the depletion layer thickness, σ the injection conductance which is essentially the same as $-G$, and θ_d the drift angle given by $\theta_d = \omega W/v$, where v is the velocity of the electrons. The total impedance of an RTD is the sum of the impedance of the quantum well region and Z_{tt}. If the transit time is dominant in the delay components of an RTD, f_{MAX} is given by the following [10]:

$$f_{MAX} = \frac{1}{2\pi}\frac{v}{2W}\left[\pi + \left\{\pi^2 - \frac{8|\sigma|W}{\epsilon v}\right\}^{\frac{1}{2}}\right] \tag{5.5}$$

Table 5.1. Material parameters related to the performance of RTDs.

	GaAs	$In_{0.53}Ga_{0.47}As$	InAs
Peak electron velocity (cm s^{-1})	2×10^7	3.5×10^7	5×10^7
Specific contact resistance (Ω cm^2)	5×10^{-6}	3×10^{-8}	5×10^{-9}

It is clear from this equation that W should be small in order to achieve a high f_{MAX}. Thus, there is an optimum value for W which compromises the reduction in C_D and the reduction in transit time delay. To overcome this trade-off, one has either to increase v or reduce the C_D-related time constants: R_sC_D and C_D/G. The reduction in C_D/G has been discussed above; the reduction in R_s and the increase in v are discussed next.

To a great extent, R_s and v are determined by the material used for the emitter and collector. If a GaAs/AlGaAs material is considered, typical ohmic metals to n^+-GaAs layers have a specific contact resistivity of about 10^{-6} Ωcm^2, and the saturation velocity of electrons in n^+-GaAs layers is less than 1×10^7 cm s^{-1}. Assuming again a specific C_D of 100 nF cm^{-2}, corresponding to a depletion layer thickness, W, of 100 nm, R_sC_D becomes 0.1 ps, while the transit time exceeds 1 ps. Thus a compromise will have to be reached between the two time constants. By changing the material to InGaAs/AlAs or InAs/AlSb, however, a specific contact resistivity less than 10^{-7} Ωcm^2 and a saturation velocity more than 1.5×10^7 cm s^{-1} can be achieved [11], [12]. This is a significant improvement compared with the GaAs/AlAs system. Typical material-related parameters reported to date are summarised in Table 5.1.

A comparison of the high-speed performance of RTDs has been conducted by Brown [2] for different material systems, as shown in Fig. 5.3. The highest oscillation frequency of 712 GHz was reported [1] for an InAs/AlSb double-barrier RTD. As a result of the superb material characteristics, the InAs/AlSb RTD exhibits both the highest power and highest frequency operation.

There does not seem to be much scope left for further improvement in the high-frequency performance of RTD oscillators since the structure of RTDs is very simple. There is, however, some possibility, especially for RTDs with GaAs collectors, that the ballistic collection scheme [13] may reduce the transit time of electrons across the depletion layer in the collector. The ballistic collection scheme controls the electric field in the

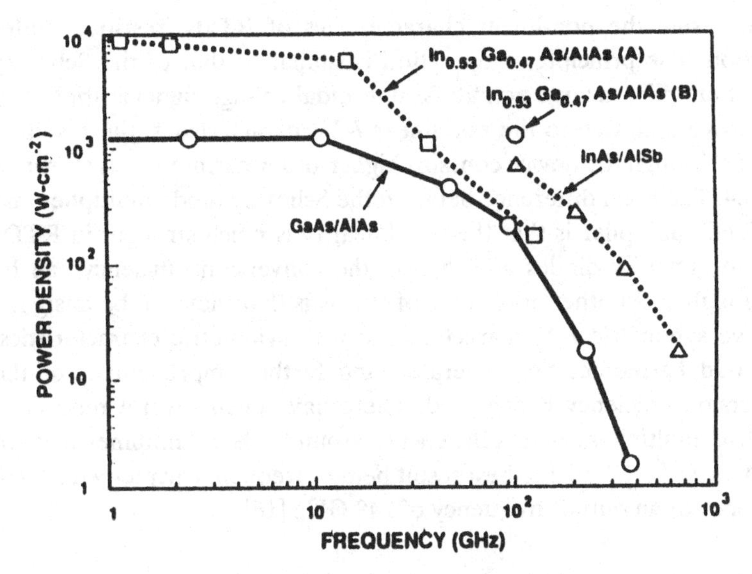

Figure 5.3 Experimental performance of RTD-oscillators as a function of frequency for three different materials [4]. Reprinted with permission of MIT Lincoln Laboratory, Lexington, Massachusetts.

collector by tailoring the doping profile so that the kinetic energy of the electrons does not exceed the energy separation between the Γ-valley and the X- or L-valley over the distance of the depletion layer. In this manner, conduction electrons in the ballistic collection structure stay in the Γ-valley while having the largest possible velocity overshoot. As a whole, these electrons have a higher average velocity than do those in the conventional collector structure in which they are frequently scattered into the X- or L-valleys at higher electric fields. As the energy separations between the Γ- and X- or L-valleys are as small as 0.3 eV, this scheme has proved to be very effective in GaAs and has been adopted in the collectors of heterojunction bipolar transistors. However, even in InGaAs in which the Γ–X- or Γ–L-separations are much larger, a high electric field in the collector has been found to increase the transit time. Therefore, it is also possible for this ballistic collection scheme to work for RTD collectors, loosening to some extent the trade-off between C_D and the transit time.

So far in this section, the properties of RTD oscillators have been described. There is, however, another way to create high-frequency signals using the non-linear characteristics of RTDs: resistive multiplication. The principle of operation is similar to that of the Schottky diode multiplier: a large amplitude sinusoidal voltage signal is applied to the device and, due to the non-linear *I–V* characteristics, the resulting current through the device contains higher-order harmonics, as shown in Fig. 5.4. The main difference between the Schottky diode multiplier and the RTD multiplier is that the non-linearity is much stronger in RTDs than in Schottky diodes and, hence, the conversion efficiency can be much higher. Another advantage of RTDs is that they can be designed to have symmetric *I–V* characteristics: with symmetric characteristics, only odd harmonics are generated and further improvement of the conversion efficiency is achieved. There have been several reports of resistive multipliers with efficiencies around 1% at millimetre wave frequencies [14]–[16], the best result being a frequency tripler with 1.2% efficiency at an output frequency of 249 GHz [16].

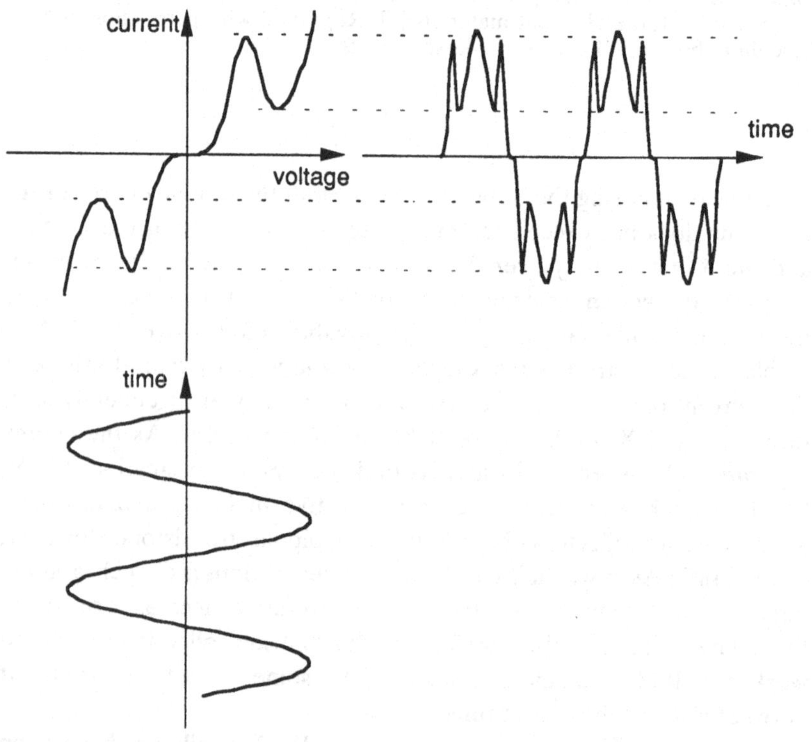

Figure 5.4 Schematic illustration of frequency multiplication using the *I–V* curves of an RTD.

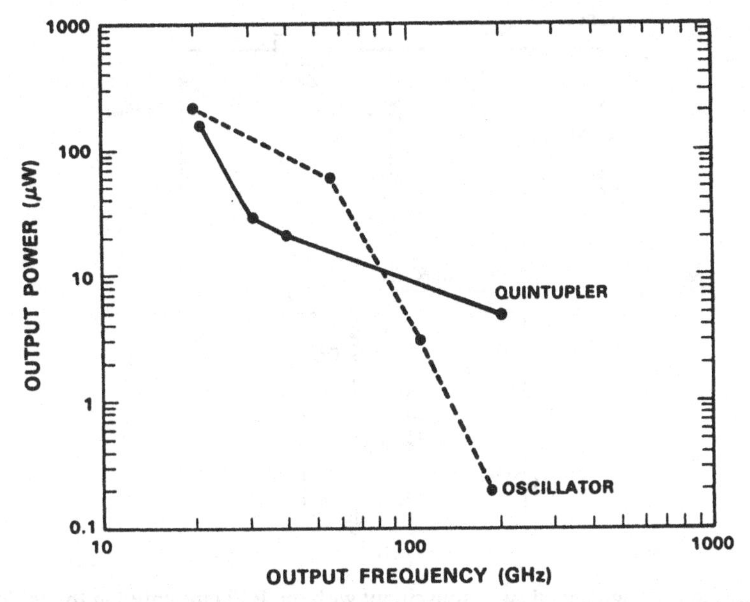

Figure 5.5 Output power as a function of frequency for RTD fundamental circuit oscillators and RTD multipliers [3]. Reprinted with permission of MIT Lincoln Laboratory, Lexington, Massachusetts.

One of the advantages of the multiplier approach over a fundamental oscillator is that the output power of the multiplier can be larger than that for an oscillator at high frequencies [3], as shown in Fig. 5.5. This is because resistive multiplication does not necessarily require gain from the device, and so the frequency limit of f_{MAX} does not apply. Therefore, supposing that a high-power signal source is available and that the undesirable harmonics can be terminated successfully, a resistive multiplication circuit may serve as a more powerful signal source than a fundamental oscillator.

5.1.2 High-speed switching and its application

Because of their NDC characteristics and extremely fast response, RTDs can also be used in high-speed switching circuits [17]–[21]. The principle of operation, studied in detail in reference [18], is similar to that for a conventional switch based on a tunnel diode. Figure 5.6(a) [18] illustrates a switching circuit using an RTD as the switching element. In

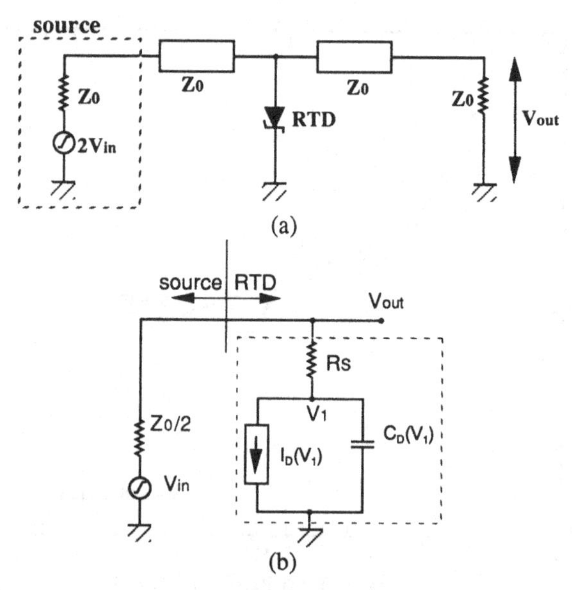

Figure 5.6 (a) High-speed switching circuit with the RTD mounted in the middle of a transmission line and (b) its equivalent circuit in which the equivalent circuit for the RTD is shown in the dashed rectangle. The NDC characteristic of the RTD is expressed as the function $I_D(V_1)$.

Fig. 5.6(b), an equivalent circuit is shown which includes the RTD equivalent circuit for large-signal analysis. This differs from Fig. 5.1 in two ways: first, a small-signal conductance $-G$ is replaced by a current function $I_D(V_1)$ as a function of the voltage across the diode, and, second, L_{QW} is not taken into account. Expressing the diode conductance as a function of voltage is necessary for large-signal analysis since the differential conductance of the diode changes sign during switching.

The current function $I_D(V_1)$ represents the I–V characteristics of the RTD and is assumed to be a piecewise-linear function, as shown in Fig. 5.7. In the following, let us consider the circuit operation and the switching behaviour of RTD switches. Using the notation of Fig. 5.6(b), the basic circuit equation in terms of current is as follows:

$$C_D(V_1)\frac{dV_1}{dt} = \frac{(V_{in} - V_1)}{R_L + R_s} - I_D(V_1) \tag{5.6}$$

where $R_L = Z_0/2$. If we write the first term on the right-hand side as

$$I_L(V_1) = (V_{in} - V_1)/(R_L + R_s) \tag{5.7}$$

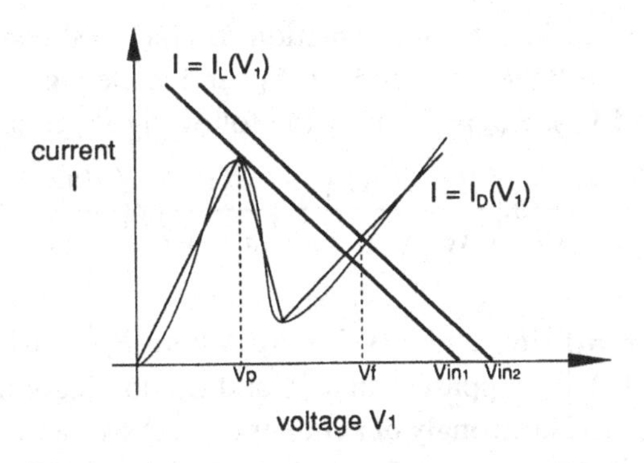

Figure 5.7 Piecewise-linear *I–V* characteristics, $I_D(V)$, for an RTD.

because it is the current supplied by the source, eqn (5.6) is reduced to

$$C_D(V_1)\frac{dV_1}{dt} = I_L(V_1) - I_D(V_1) \tag{5.8}$$

The steady-state solutions can be obtained by neglecting the time derivative and solving

$$I_L(V_1) = I_D(V_1) \tag{5.9}$$

Though $I_D(V_1)$ is a non-linear function, eqn (5.9) can be solved graphically by plotting lines corresponding to $I = I_L(V_1)$, which is usually called the load line, and $I = I_D(V_1)$. The solutions are given by the intersection of the two lines. In order to study the switching operation, two cases are shown in Fig. 5.7. In the first case, the input voltage, V_{in1}, is chosen so that the two lines intersect at the current peak of the RTD, giving two solutions: one at $V_1 = V_p$ and the other at a higher voltage. The other case, which has a slightly higher input voltage of V_{in2} has only one solution at $V_1 = V_f$. Supposing the initial state is $V_{in} = V_{in1}$ with $V_1 = V_p$, and that the input voltage is suddenly changed to V_{in2}: the diode voltage switches from V_p to V_f. Note that the negative slope of NDC, $1/R_n$ in Fig. 5.7, must be steeper than the slope of $I = I_L(V_1)$, that is, $1/(R_L + R_s)$, to allow this switching.

The transient behaviour of the circuit can be expressed through eqn (5.8). By rewriting this to $dt = dV_1[C_D(V_1)/\{I_L(V_1) - I_D(V_1)\}]$ and integrating from $V_p + 0.1\Delta V$ to $V_f - 0.1\Delta V$, where $\Delta V = V_f - V_p$, the 10–90% rise time, T_{rise}, is given by

$$T_{rise} = \int_{V_p+0.1\Delta V}^{V_f-0.1\Delta V} \frac{C_D(V)}{I_L(V) - I_D(V)}\, dV \tag{5.10}$$

Using the piecewise-linear approximation for $I_D(V_1)$ shown in Fig. 5.7, and assuming a voltage-independent C_D and a negligible difference between V_{in1} and V_{in2}, T_{rise} is given by the following expression [18]:

$$T_{rise} = |R_n|C_D\left\{\left(\frac{x}{x-1}\right)\ln\left(\frac{10(x+y)}{x(x-y)}\right) + \left(\frac{xy}{x-y}\right)\ln\left(\frac{10(x+y)y}{x(x-y)}\right)\right\} \tag{5.11}$$

where $x \equiv (R_L + R_s) / |R_n|$ and $y \equiv R_d / |R_n|$, where $R_d^{-1} \equiv dI_D V|_{V-V_f}$.

From eqn (5.11), it is apparent that R_n and C_D dominate the switching speed. And T_{rise} is also strongly dependent on x. As x decreases towards one, which is its minimum, T_{rise} increases since $I_L(V) - I_D(V)$, the current that charges up the diode capacitance, becomes smaller. However, if x becomes large, due to an increased load impedance, the voltage swing required for switching also rises resulting in an increase in the switching time. Therefore, there is an optimum value for x depending on the value of y. If we compare these speed-limiting factors with those of RTD-oscillators, a reduction in diode capacitance and an increase in NDC (i.e. $1/R_n$ for switching analysis) are needed for the improvement of speed in both cases. The reduction of R_s, however, has a slightly different meaning. For a switching operation, as shown above, an increase in R_s does not necessarily result in speed degradation. It does, however, decrease the output voltage swing of the circuit since the output voltage of the switch is given by multiplying $R_s/(R_s + R_L)$ and the voltage swing of the diode. In any case, for better performance, a reduction in R_s, R_n and C_D must be pursued: the design considerations are thus almost the same for both applications.

There have been some reports on the measurement of the switching time of RTDs. For a GaAs/AlAs RTD, a switching time as short as 2 ps has been reported by using photoconductive switch excitation and an electro-optic sampling technique [17]. However, the theoretical prediction for the system indicates a switching time of around 10 ps for the RTD: this is significantly longer than the measured value. It has been suggested [22] that the experiment was conducted under an overdrive condition, resulting in the discrepancy. Therefore, the limit of switching in this material system is yet to be clarified. By changing the material, as has been done with RTD oscillators, a switching time as short as 1.7 ps has been reported for an InAs/AlSb RTD. In this case, the RTD was driven by a sine wave to avoid the overdrive situation, and the switching waveform was measured by electro-optic sampling.

In the field of application of RTD switches, there have been a few

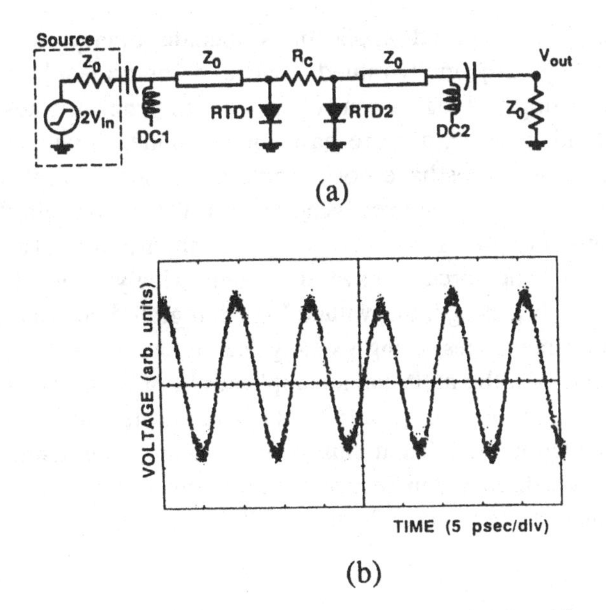

Figure 5.8 (a) An improved high-speed switching circuit with two RTDs mounted in the middle of a transmission line linked by a resistor R_C and (b) a trace of the 110 GHz signal monitored using the circuit shown in (a). After Özbay and Bloom [21], with permission.

reports on their use in triggering circuits [20], [21]. Using the circuit shown in Fig. 5.6(a) with a GaAs/AlAs RTD, a rise time of 6 ps and triggering frequency of 60 GHz have been achieved [20]. Further improvement in the circuit configuration, illustrated in Fig. 5.8, has been achieved using monolithic integration of two RTDs, resistors and co-planar transmission lines [21]. Two RTDs are used in this circuit to adjust the triggering timing to the maximum slope of the incoming signal. This reduces timing jitter compared to a single-diode triggering circuit in which triggering takes place at the maximum of the signal where the slope is a minimum. By reducing the timing jitter, a triggering operation with a frequency of 110 GHz has been confirmed, which is the highest value ever reported for a triggering circuit.

5.2 Functional applications of RTDs

Let us move on to the functional applications of RTDs. Because of their extremely non-linear *I–V* characteristics, various circuits have been proposed which would use RTDs and related devices to perform new functions. The proposed applications include binary inverters and memories [23]–[26], multi-valued logic and memories [27]–[32] and signal processing and logic circuits [33]–[36]. In general, a new technology can hold its ground more easily in an undeveloped field where preceding technologies have not become established. From this standpoint, multi-valued signal processing using RTDs with multiple current peaks seems very promising because, despite their potential advantages, multi-valued logic circuits have not been widely used due to the difficulty of fabricating multi-valued logic integrated circuits, especially multi-valued memories, using existing technologies. Thus, in this section, discussion of the multi-valued applications of RTDs is emphasised. Firstly, the operating principles of RTD memories and inverters are studied in Section 5.2.1. Then multi-valued memories utilising multiple-well RTDs are discussed in Section 5.2.2. Finally, other signal processing applications are reviewed in Section 5.2.3.

5.2.1 Static memory operation of RTDs and related devices

Using a tunnel diode or RTD, with its NDC characteristics, it is possible to construct an inverter circuit with hysteresis which can be used as a static memory element with a greatly reduced number of devices compared to conventional circuits [23], [37], [38]. Let us first consider the operation of these memory circuits. The simplest configuration consists of an RTD and a resistor, constituting a memory element as shown in Fig. 5.9(a). The operation of the RTD static memory is very similar to that of the RTD switch described in Section 5.1.2 and thus only a brief additional explanation is provided here.

As described in Section 5.1.2, a steady-state solution is available at the intersection of the *I–V* characteristics for the diode $I = I_D(V)$ and the load line $I = I_L(V)$ in the circuit shown in Fig. 5.9(a). Figure 5.9(b) shows the piecewise-linear model for $I = I_D(V)$ with three linear pieces, numbered 1, 2 and 3. Also shown are three load lines, A, B and C, for different source voltages, V_{in}. There is only one solution for lines A and C whereas three solutions, corresponding to the three linear pieces 1, 2 and 3, are possible for load line B. Let us study the stability of the solutions. The basic equation, eqn (5.8), can be rewritten with explicit expressions for $I_D(V)$ and $I_L(V)$ as

$$C_D \frac{dV}{dt} = \frac{V_{in} - V}{R_L} - \left(\frac{V}{R_i} + I_i\right) \tag{5.12}$$

where R_i is the differential resistance of the diode at intersection i ($i = 1,2,3$) and I_i is the current axis intersect of each piecewise-linear function of $I_D(V)$. Note that the voltage dependence of the diode capacitance is neglected in eqn (5.12). The solution of eqn (5.12) is expressed as

$$V_{out} = \text{Const} \cdot \exp\left\{-\frac{1}{C_D}\left(\frac{1}{R_L} + \frac{1}{R_i}\right)t\right\} + \frac{R_L \cdot R_i}{R_L + R_i}\left(\frac{V_{in}}{R_L} - I_i\right) \quad (5.13)$$

The stability of this is determined by the sign of the exponent $1/R_L + 1/R_i$: for large t the exponential function tends to zero and the solution is stable if the sign is positive, whereas the exponential function tends to infinity if the sign is negative. In the latter case, any small change in voltage or current around the steady-state solution increases exponentially with time: the solution is therefore unstable. In the case of Fig. 5.9(b), 1 and 3 are stable solutions; however, solution 2 is unstable

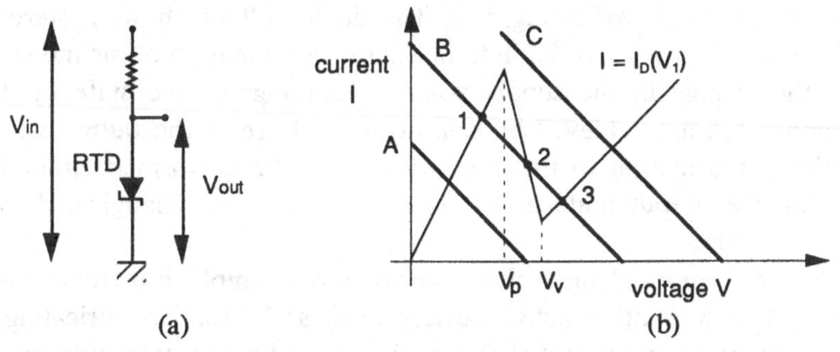

Figure 5.9 (a) A simple memory circuit comprised of an RTD and a resistor. (b) Piecewise linear modelling of the I–V characteristics for an RTD and load lines indicated by A, B and C for three different source voltages. In the case of B the circuit has two stable points, 1 and 3, and one unstable point, 2.

since $1/R_L + 1/R_i$ must be negative in order for the two curves in Fig. 5.9(b) to have three intersection points. Hence, for load line B, there are two stable points, 1 and 3. By changing the states between these two stable points, high or low, this circuit can thus function as a static memory to store information.

The operation of the static memory is as follows. In the circuit shown in Fig. 5.9(a), if V_{in} increases from zero, the diode voltage changes along the first linear part of $I_D(V)$, from zero through point 1 until it reaches V_p. Then, with a further increase in V_{in}, the diode voltage jumps, or switches, to the third linear part of $I_D(V)$, as discussed in Section 5.1.2.

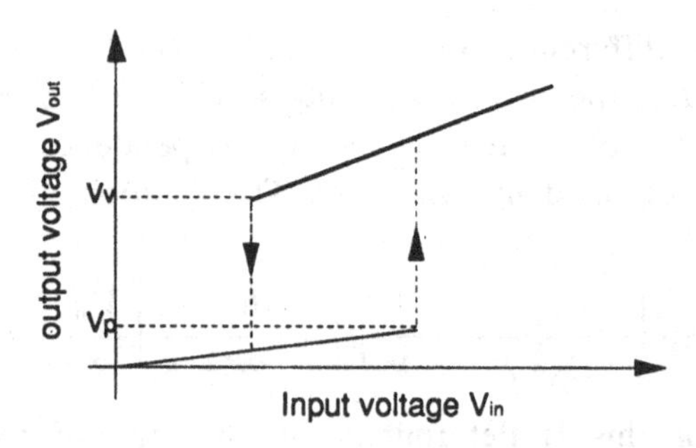

Figure 5.10 Input–output characteristics of the circuit shown in Fig. 5.9.

When V_{in} decreases after the voltage jump, the diode voltage decreases on the same line, through point 3 until it reaches V_v. A further decrease in V_{in} causes another step change in the diode voltage, this time to the first linear part of $I_d(V)$. Therefore, plotting the diode voltage, or the output voltage, as a function of the source voltage, V_{in}, results in the characteristics shown in Fig. 5.10. The diode voltage shows hysteresis whose width in terms of V_{in} determines the noise margin of the memory for the change in the supply voltage. Readings of the state of the memory cell are achieved by sensing the voltage at the output node. Writing information to the memory cell can be performed either by pulling the output node up or down directly or by changing V_{in}, as discussed above.

As the memory element shown above is very simple, it is suitable for integration with other active devices, such as FETs, for fabricating a memory integrated circuit (IC). For this type of RTD static memory, a typical circuit configuration would be as shown in Fig. 5.11.

In this configuration, memory elements are placed in a matrix array, and a cell in the array is addressed by selecting a particular column and row, which opens gates Q_1 and Q_2, connecting the output node of the memory element to the bit line. Then, a reading operation can be performed by sensing the node voltage, while a writing operation can be performed by forcing the node voltage to be high or low through the bit line. Compared to the conventional static memory cell, which needs six transistors or four transistors and two resistors for each memory cell, the RTD memory cell needs only three devices: an RTD, a resistor and an FET. Furthermore, the RTD memory can operate without the complex internal feedback interconnections inside each memory cell which are indispensable in conventional circuits. Therefore, the RTD memory cell occupies much less area than can be provided in conventional static

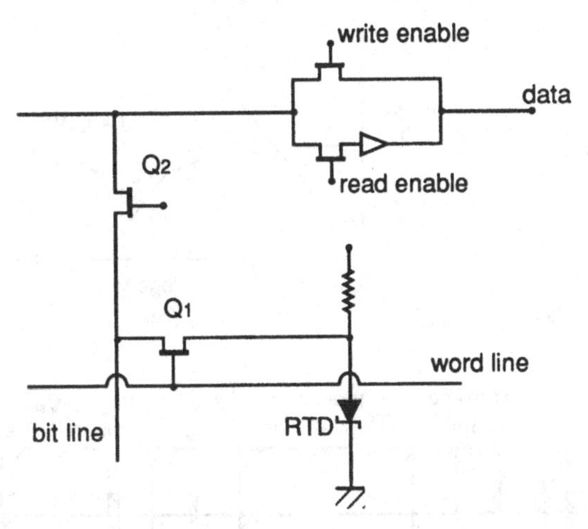

Figure 5.11 A typical configuration of an RTD static memory IC.

memories, using current microfabrication technologies, and is suitable for large-scale integration.

In designing and fabricating RTD memory elements, the most crucial factor is the reproducibility of the *I–V* characteristics, especially the valley current, of the RTDs. As can clearly be seen from Fig. 5.9(b), an increase in the valley current can extinguish intersection point 3 which is one of the two stable points of the memory operation. To avoid this failure, the load resistance, $R_{l'}$, and consequently the supply voltage, must be chosen to be high. This increases the required amplitude of the writing pulse, leading to a significant increase in power dissipation. In Chapter 2 it was shown that a simple theoretical prediction based on the global coherent tunnelling model is fairly accurate for the peak voltage and the peak current density but is not so accurate for the valley current and the valley voltage. We saw in Chapters 3 and 4 that the valley current is predominantly determined by phase-coherence breaking scattering in the quantum well which, at present, is yet to be controlled. In view of the requirements of circuit design and fabrication mentioned above, however, it is desired to find a microscopic way of controlling these scattering events in order to reduce the valley current.

Different configurations are possible for the RTD memory element by replacing the resistive load by other non-linear elements. Figure 5.12(a) shows the circuit for an RTD memory element with a depletion mode FET (D-FET) load. The gate electrode of the D-FET is either shorted to the source electrode or is driven externally. In the former

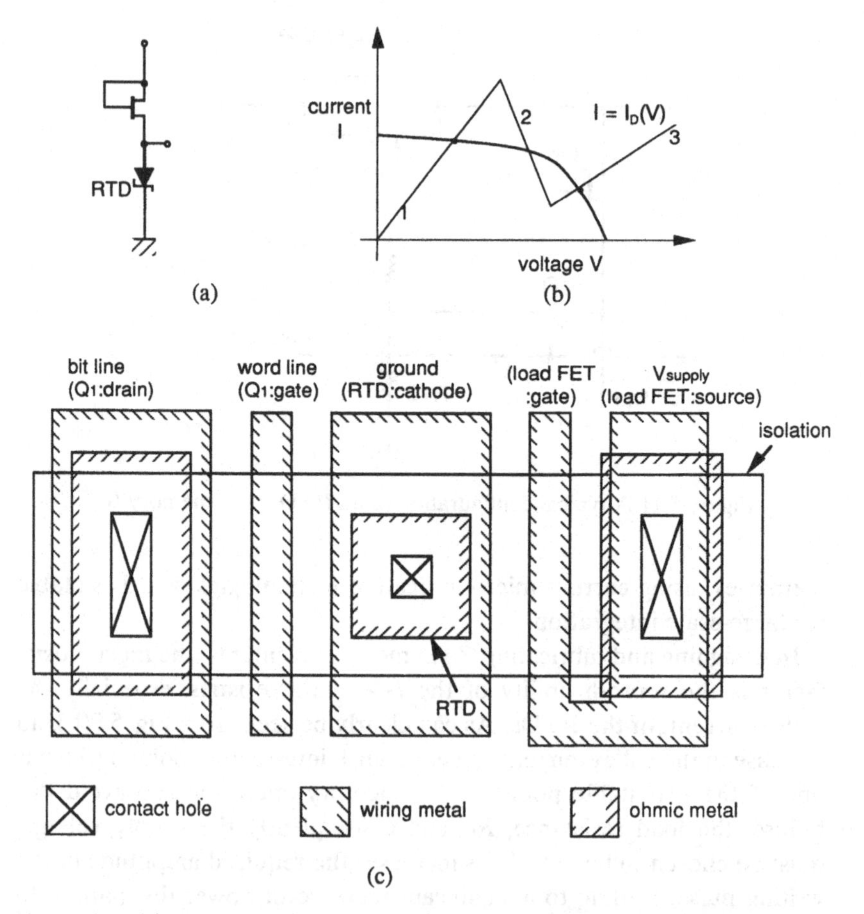

Figure 5.12 (a) A circuit of an RTD memory element with a depletion mode FET (D-FET) load. (b) Piecewise linear modelling of the *I–V* characteristics for an RTD and D-FET load. (c) A layout example of a static memory cell utilising an RTD with a D-FET load.

configuration, the load line becomes as shown in Fig. 5.12(b). The D-FET load can be regarded as a constant current source in the current saturation region because of its very high differential output resistance. Thus, charging and discharging of the RTD capacitance can be quicker than using a resistive load in which charging current decreases with diode voltage. The operating margin for any supply voltage fluctuations is also increased by using the D-FET load because of its high output resistance. To achieve a high load resistance with a resistive load, the power supply voltage must be very high; it should be higher than the product of the load resistance and the valley current. In contrast, the D-FET load enables us to use a lower power supply voltage than does the resistive load since, as shown in Fig. 5.12(b), the FET load line bends towards the voltage axis near the supply voltage. Furthermore, by setting the power supply voltage slightly above the valley voltage, the holding current of the high state can be reduced to the valley current [23], [24], [37], thus reducing the power consumption. In the resistive load circuit, setting the high state current to the valley current can be realised only by sacrificing the operating margin to null since a very small decrease in the supply voltage or a small increase in the valley current can delete the operation point. With these advantages the D-FET load is much more attractive than the resistive load.

An additional advantage of the FET load memory element is that it is suitable for integration. For example, a memory cell with a transfer gate Q_1, as shown in Fig. 5.11, can be integrated as shown in Fig. 5.12(c) if the RTD structure is formed directly on top of the n^+ source or drain [23], [39], [40]. In this layout, the load FET, transfer gate Q_1, and the RTD are connected not by a metal interconnection but in the highly doped semiconductor layer. Since no contact pads are necessary, the cell area can be reduced to less than that of two separate FETs.

The circuits described so far can be operated only as memory cell elements. However, if the gate of the FET load is not connected to its source but is controlled from outside, the circuit becomes an inverter with hysteresis in input–output characteristics [23], [24]. In this configuration, the RTD is usually regarded as a load and the FET as a driver [23]; however, to maintain consistency with the previous discussion, the FET is regarded as a load element throughout this chapter. The mechanism of operation is illustrated in Fig. 5.13. By changing the gate

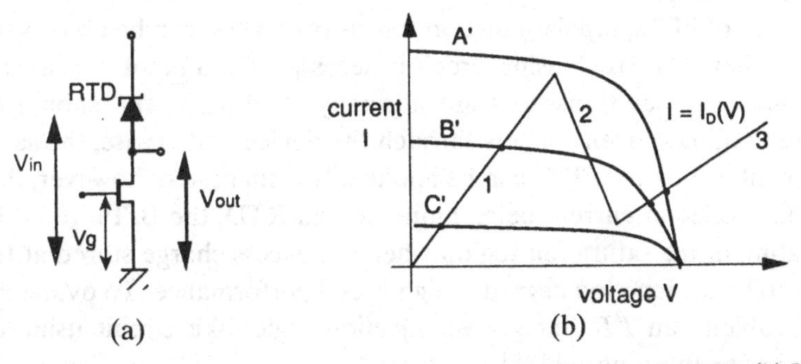

Figure 5.13 (a) An inverter circuit comprising an RTD with a D-FET. (b) Its *I–V* characteristics for three values of the gate bias of the FET.

bias, Vg, of the load FET, the saturation current of the FET can be changed as shown by load lines A', B' and C' in Fig. 5.13, with the number of stable operation points being 1, 2 and 1 for each load line respectively. The input–output characteristics of the inverter shows hysteresis, as illustrated in Fig. 5.14. If the gate voltage of the FET is

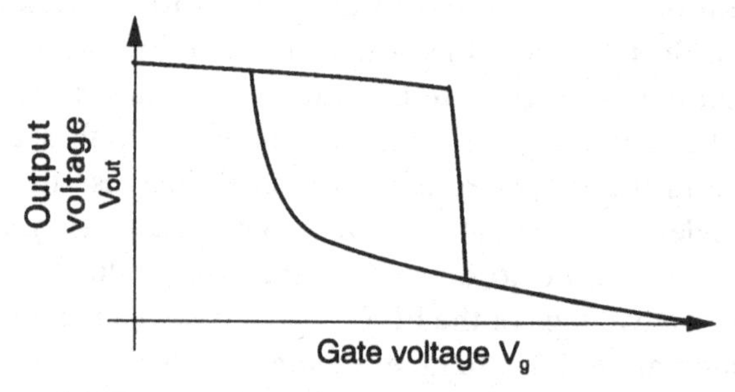

Figure 5.14 Input–output characteristics of the RTD–FET inverter.

retained within the hysteresis regime, the inverter can be operated as a memory element, and the state of the memory can be changed by controlling the gate voltage of the FET. Because the FET incorporates gain, it takes less external power to control the gate voltage than it does to change the voltage of the output node of the memory element directly in order to write information into the memory cell; however, to write into the cell by controlling the gate of the load FET requires one more interconnection in addition to the reading interconnection shown in Fig. 5.11 between the external circuits and the memory cells. Thus, this configuration has a certain advantage as a memory element over that shown in Fig. 5.12 if a reduction in power consumption for writing the data into the memory cell is required, but at the expense of a reduced packing density.

Instead of FETs, bipolar junction transistors (BJTs) can be used as the load; however, a base biasing circuit is necessary for a bipolar transistor load since bipolar transistors are normally-off devices and should be biased to allow current to flow through the device. Otherwise, the same argument as that for FETs can be applied. It is suggested, however, that with the collector current being limited by an RTD, the BJTs would be operating in the saturation region where an excess charge stored at the base–collector junction degrades high-speed performance. To overcome this problem, an I^2L (integrated injection logic)-like circuit using an RTD has been proposed [25].

Figure 5.15(a) shows another variation of an RTD memory element in

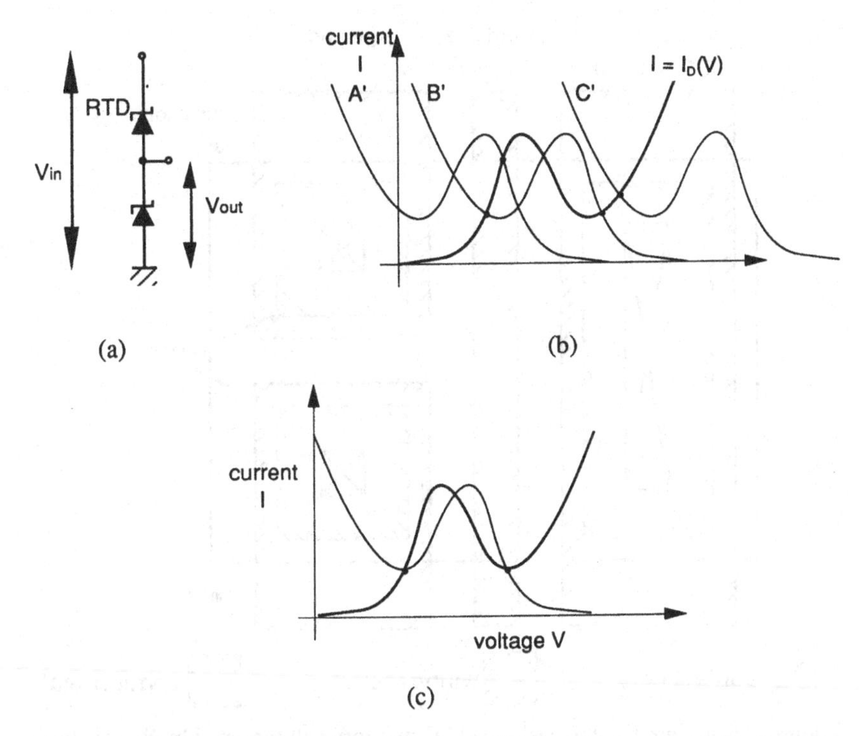

Figure 5.15 (a) A cascade RTD memory cell, (b) its *I–V* characteristics for three different values of V_{in}, and (c) a configuration with the lowest power consumption.

which the load resistor is replaced by a second RTD, forming a cascade of two RTDs [25], [26]. In this circuit, the load line becomes more complex than that for a D-FET load or resistive load; however, the essential operation of the circuit remains the same. In Fig. 5.15(b), a typical load line and the diode current as a function of diode voltage are shown for different supply voltages. The two *I–V* curves have a maximum of three points of intersection; however, only two points are stable, for exactly the same reason as before.

The advantage of this configuration is that, by adjusting the supply voltage, each stable point can be designed to take place at the valley of the *I–V* characteristics of each RTD, as shown in Fig. 5.15(c), minimising the current required to hold the state of the memory. This is a unique feature of this cascade RTD memory cell. In a resistive load memory element or a D-FET load cell, a small operating margin is normally allowed so that the stable point is not at the minimum current level,

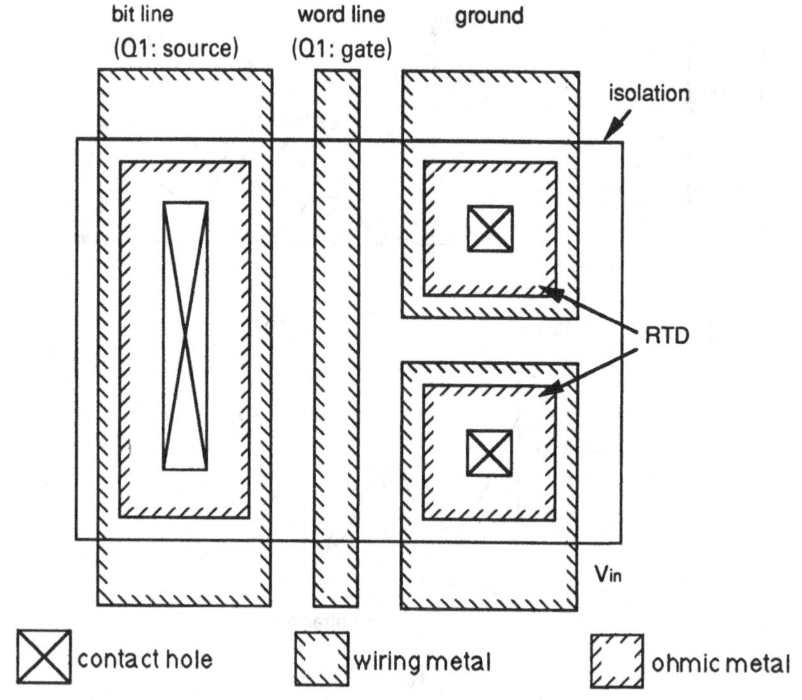

Figure 5.16 A layout of the cascade RTD memory cell reported by Watanabe *et al.* [25].

since otherwise a small decrease in the supply voltage would change the state of the memory from the high to the low state, as discussed previously. In contrast, this cascade RTD memory cell has a larger operating margin, since the load line shown in Fig. 5.15(c) crosses the current valley with a steeper slope than would the usual resistive load line due to the shape of the *I–V* curve of RTDs; therefore, this cascade RTD memory element is suitable for a memory element with reduced power consumption.

This cascade RTD memory element is also preferable in terms of integration. Watanabe *et al.* have reported a memory cell including a transfer gate Q_1 as shown in Fig. 5.16 [26]. In this layout, the two RTDs are fabricated on top of the drain of the transfer gate Q_1, reducing the cell size to only slightly more than the size of one FET. The cell size was 12×18 μm, less than half that for a typical FET SRAM, and the average power dissipation was 3 μW with a supply voltage of 1V.

Before closing Section 5.2.1, it is worth introducing a new resonant tunnelling structure which has been reported very recently by Gullapalli *et al.* as a promising candidate for a future memory device [41]. This RTD consists of a double-barrier structure and a special N^-–N^+–N^- doping profile in the spacer layers, as depicted in Fig. 5.17(a). The *I–V* characteristics exhibit a unique multi-stable behaviour with various

1770 ML	N+ GaAs	4×10^{18} cm^{-3}
65 ML	N- GaAs	10^{15} cm^{-3}
47 ML	N+ GaAs	4×10^{18} cm^{-3}
18 ML	N- GaAs	10^{15} cm^{-3}
6 ML	Unintentionally Doped AlAs	
18 ML	N- GaAs	10^{15} cm^{-3}
6 ML	Unintentionally Doped AlAs	
18 ML	N- GaAs	10^{15} cm^{-3}
47 ML	N+ GaAs	4×10^{18} cm^{-3}
65 ML	N- GaAs	10^{15} cm^{-3}
	n+ GaAs Substrate	

(a)

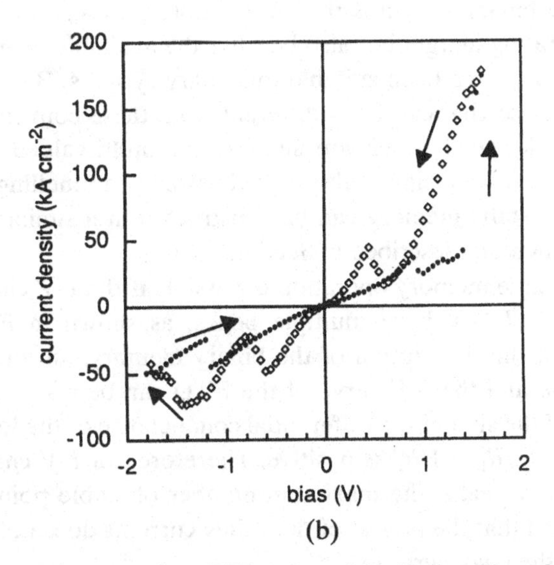

(b)

Figure 5.17 (a) Schematic of an RTD reported by Gullapalli *et al.* and (b) *I–V* characteristics observed at room temperature. After Gullapalli *et al.*, private communications.

values of resistance even at zero bias, as shown in Fig. 5.17(b). It has been reported that, by applying a large bias, switching between the two states can be induced; no spontaneous transition is, however, observed, even after the two terminals are shorted for several minutes. It is suggested from Gullapalli *et al.*'s self-consistent calculations, similar to those described in Chapter 2, that there are four stable states at zero bias, although it is unclear which two of the four states have been observed experimentally. Though there is still much to be considered, such as the stability of the states, this structure has considerable potential for use as an element in a non-volatile memory: a system that has not yet been achieved in compound semiconductor devices.

5.2.2 Multiple-well resonant tunnelling structures and multi-valued memories

This section deals with more complex resonant tunnelling diodes with multiple quantum wells (see also Section 2.5) and their applications to multi-valued memory. So far in the semiconductor industry, binary logic circuits and memories have prevailed, even though multi-valued logic has been said to have many advantages over binary systems, such as a reduced number of stages for certain logic functions. This is not only because the binary system is the most simple to design and allows the largest operating margin but also because the available devices, such as FETs or BJTs, have been suitable for binary systems. By using RTDs, however, device characteristics different from these conventional devices can be designed which are suitable for multi-valued systems. In particular, by introducing multiple-well resonant tunnelling structures, multi-valued static memory can be constructed in a similar manner to the binary memory described in Section 5.2.1.

Multiple-state memory operation is possible if the I–V characteristics for an RTD, $I_D(V)$, have multiple peaks, as shown in Fig. 5.18. By analogy with the description of the binary memory, an intersection of the load line and the I–V curve of the RTD can be a stable operation point only if the sum of the differential conductance of the load and that of the diode ($1/R_L + 1/R_i$) is positive. Therefore, for I–V characteristics with n current peaks, the maximum number of stable points is n plus one, provided that the largest of the valley currents does not exceed the smallest of the peak currents.

With a double-barrier RTD, it is possible to observe the second or third current peaks arising from the higher resonances of the second or

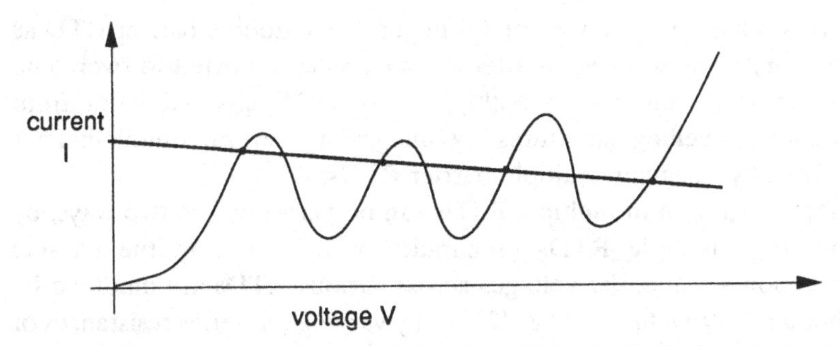

Figure 5.18 Schematic *I–V* characteristics of an RTD suitable for a multiple-state memory application.

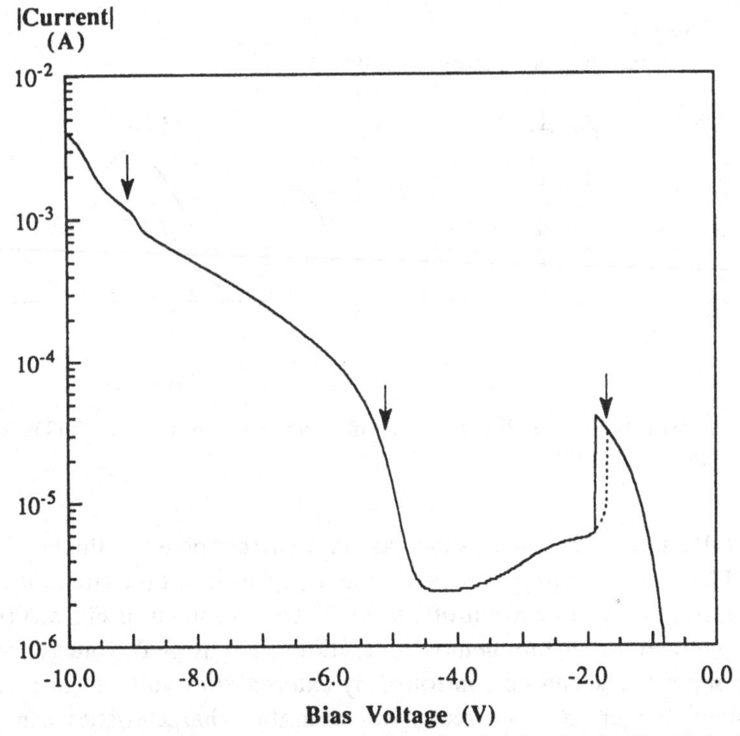

Figure 5.19 *I–V* characteristics for Material 2 (see Section 2.4 for detail) for large applied voltages. Arrows show predicted structures due to higher resonances. After Goodings [42], with permission.

third quantised states in the quantum well for voltages above the first current peak (see Fig. 5.19 [42]); however, due to the high excess current at higher voltages, valley currents for the second or higher resonances usually exceed the first peak current. Therefore, multiple-state memory operation has not been reported using only one double-barrier RTD as a memory element. Several attempts have been reported to overcome this limitation and create multiple peak NDC characteristics from resonant tunnelling structures by integrating several double-barrier RTDs or by designing multiple-barrier RTDs.

The integration of multiple RTDs can be performed in two ways: by connecting multiple RTDs in parallel or in series. In the parallel connection scheme, the voltages across various RTDs are modified by either an external bias voltage [27] or by varying the series resistances of the diodes [30], shifting the peak voltages of the diodes with respect to each other, as shown in Fig. 5.20(a).

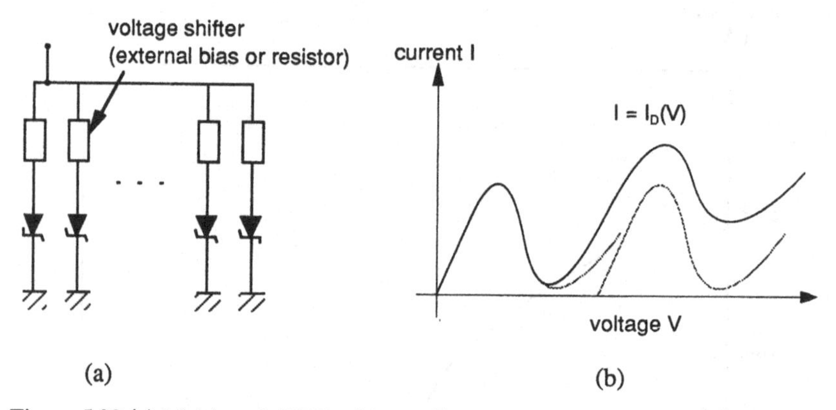

Figure 5.20 (a) Multi-peak RTD with parallel connection of multiple RTDs, and (b) its operational principle.

The *I–V* curve of the circuit shows as many current peaks as the number of RTDs connected in parallel since the total current of the circuit is the superposition of the current of all the RTDs, as shown in Fig. 5.20(b). Key parameters for the memory operation, such as peak voltages and their separations, can be controlled by external bias voltages or series resistances; therefore, a circuit with desirable characteristics can be designed. However, the valley current tends to be much larger than that of a single diode since the currents of all the RTDs connected in parallel are superimposed. Furthermore, the total area of the circuit increases in proportion to the number of diodes; it becomes n times that of a single diode if n peaks are necessary since each RTD contributes to a single current peak.

In order to avoid the above-mentioned disadvantages, another scheme which connects multiple RTDs in series was proposed. The operation of this circuit is explained below, where we consider the simplest case in which two diodes are connected in series. At a low supply voltage, the operation points of the series RTDs are at the lowest branch of the *I–V* curves of both diodes, as shown in Fig. 5.21(a). By increasing the bias voltage, the current through the diodes increases and, eventually, reaches the smaller of the peak currents of the two diodes connected in series, as shown in Fig. 5.21(b). As the bias is increased

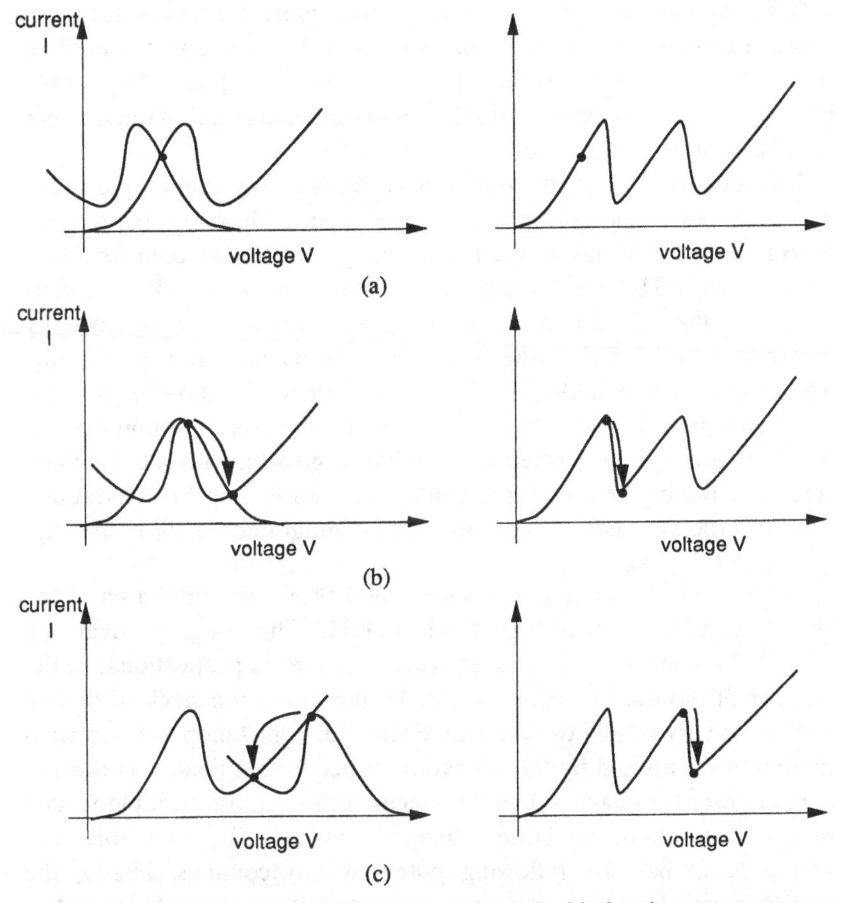

Figure 5.21 Operation principle of a stacked double-barrier structure as a multiple-peak NDC diode. On the left-hand side, the single-diode characteristics with a load line are illustrated whereas, on the right, the *I–V* curve of the total stack is shown: (a) corresponds to a bias lower than both NDC peaks, (b) corresponds to a bias at the first peak-to-valley transition, (c) corresponds to a bias at the second peak-to-valley transition.

further, the voltage across the diode with the smaller peak current changes from its peak voltage to a higher value, as indicated by the arrow in Fig. 5.21(b). Thus, the current through the circuit shows the first transition from a peak to a valley. When the bias is increased further, the current through the circuit increases again and reaches the peak current of the second diode, as shown in Fig. 5.21(c). Further increase of the supply voltage causes another transition of circuit current, as indicated by the arrow in Fig. 5.21(c). Thus, two current peaks corresponding to the two transitions can be observed with two RTDs connected in series. For circuits with more RTDs connected in series, a process similar to the one described above is repeated until all the RTDs connected in series have experienced peak-to-valley transition. Therefore, the number of NDC peaks is again equal to the number of RTDs connected in series.

This idea of using multiple diodes in series to create multiple peaks was originally proposed for Esaki tunnel diodes. However, the scheme works better for RTDs, especially for integrated RTDs, than for Esaki diodes since RTDs can be designed to have symmetric *I–V* characteristics and the fabrication of RTD layer structures is performed by epitaxial growth without the need for impurity diffusion or alloying. These facts enable multiple RTDs to be connected directly in semiconductor materials; in other words, a multiple stack of double-barrier RT structures can be formed using epitaxial growth, with the complete stack functioning as a multiple peak diode. So far, up to five double-barrier structures have been successfully integrated to show five significant NDC peaks [33].

In the stacked double barriers described above, the total area of the circuit can be the same as that of a single RTD. This is a great advantage over RTDs connected in parallel, where the area is proportional to the number of diodes, or current peaks. Furthermore, the stack of double barriers is very simple to fabricate. From a design standpoint, a certain limitation is imposed by the operation principle that peak currents are always lower for peaks at lower voltages, although, for multiple-valued memory operation, this is not a major drawback. However, this integrated RTD has the following potential shortcomings. Firstly, the structure may lead to unexpected multi-stability in the *I–V* characteristics resulting from heavily doped regions whose electrical potentials are not fixed externally. Unless all the double-barrier structures are exactly the same, the system could be switched from one stable state to another with a different current density by a weak external perturbation.

Table 5.2. Structural parameters of the triple-well RTD

Layer		Thickness (nm)
i-AlAs	L_{B1}	1.0
i-GaAs	L_{W1}	11.9
i-$Al_{0.26}Ga_{0.74}As$	L_{B2}	3.0
i-GaAs	L_{W2}	5.7
i-$Al_{0.26}Ga_{0.74}As$	L_{B3}	3.0
i-GaAs	L_{W3}	6.9
i-AlAs	L_{B4}	1.0

Secondly, the heavily doped regions required to avoid undesirable quantum interference between adjacent double barriers increase the total thickness of the devices. In our experiment, a 50-nm-thick layer with 1×10^{18} cm^{-3} doping does not nullify the quantum interference between adjacent double barriers completely; therefore, the total thickness of layers for each sequence, including a double-barrier structure with undoped spacers on both sides and a heavily doped layer, will become more than 100 nm. Considering the integration of the stacked double-barrier structures with underlying FET layers, as described previously, this may impose a difficulty in processing the FETs with precise control of the threshold voltage after removing the thick RTD layers.

In the following section, another approach to creating multiple NDC using a multi-barrier diode is described (see also Section 2.5). This approach is expected to solve the above-mentioned shortcomings of the stacked double-barrier structure. The main idea is to create resonant tunnelling structures in which multiple resonances can occur resulting in the ideal multiple-NDC characteristics. In order to create multiple resonances with similar current densities, a triple-well (quadruple-barrier) resonant tunnelling diode is examined and applied to triple-valued logic operation.

Figure 5.22 shows the band diagram of the newly proposed triple-well RTD [31], [32], which consists of three quantum wells of undoped GaAs (denoted W1, W2 and W3), four barriers (two undoped AlAs and two undoped $Al_{0.26}Ga_{0.74}As$), and highly doped n-type GaAs ($N_D = 1.0 \times 10^{18}$ cm^{-3}) emitter and collector layers on each side. The structural parameters are listed in Table 5.2.

The energy-band diagram calculated self-consistently (see Section

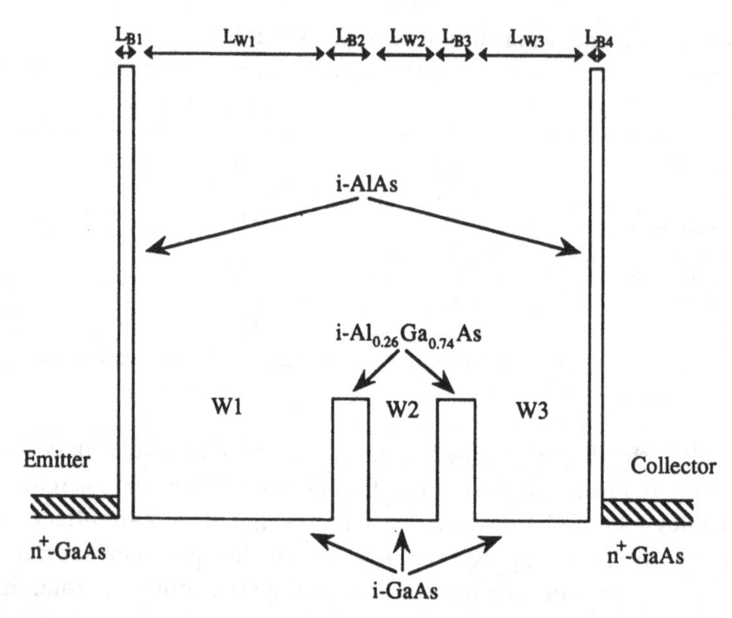

Figure 5.22 Schematic energy-band diagram of the triple-well resonant tunnelling structure.

2.2.2) near zero bias ($V = 20$ mV) is shown in Fig. 5.23(a), along with the corresponding energy dependence of the transmission probability in Fig. 5.23(b). In Fig. 5.23(a) the probability density of electrons at quasi-eigenstates, $|\Psi_R(z)|^2$, is plotted over the energy-band diagram. The calculation has been performed for energies ranging from zero to the top of the $Al_{0.26}Ga_{0.74}As$ barriers and a total of seven quasi-eigenstates has been found. The electron density distribution obtained self-consistently is shown in Fig. 5.23(c). The three lowest quasi-eigenstates (denoted Ψ_R^1, Ψ_R^2 and Ψ_R^3 hereafter) are thought to result in the double NDC in the *I–V* characteristics, as described below.

As shown in Fig. 5.23(a), $|\Psi_R^1(z)|^2$, $|\Psi_R^2(z)|^2$ and $|\Psi_R^3(z)|^2$ strongly reflect the natures of the ground states of wells W1, W3 and W2 respectively. Consequently, the quasi-eigenenergies E_R^1, E_R^2, and E_R^3 are most determined by the well thicknesses L_{W1}, L_{W3} and L_{W2}. In this structure the widest well, W1, has been designed so that the lowest quasi-eigenstate is located below the quasi-Fermi energy in the emitter so that it acts as an injection level for incoming electrons: only electrons with incident energy close to E_R^1 are transmitted into well W1. The peak electron density in W1 is thus higher than those in W2 and W3 in Fig. 5.23(c).

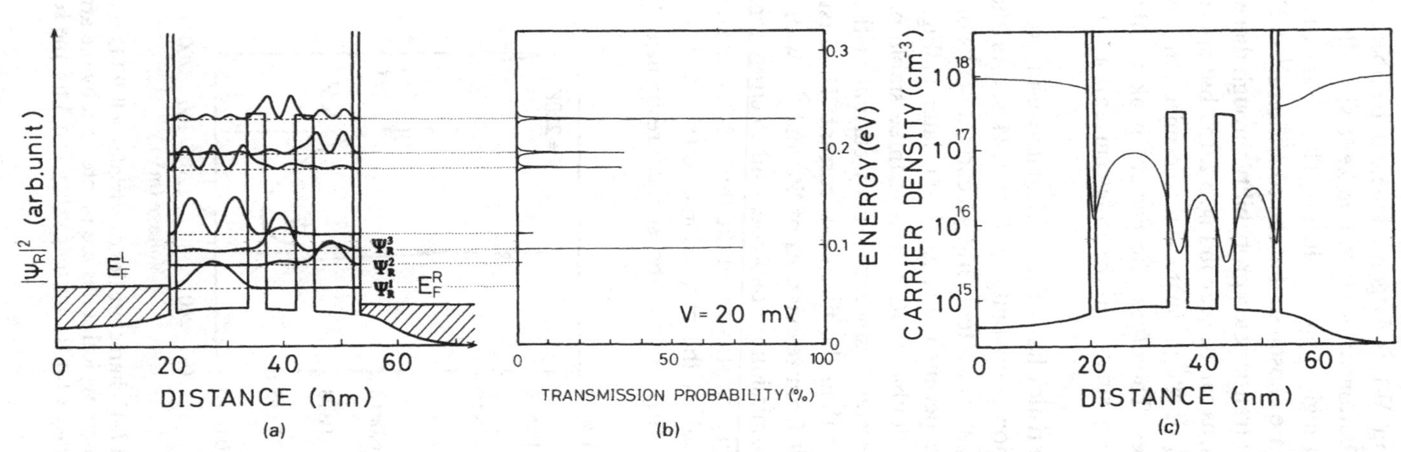

Figure 5.23 The energy-band diagram (a), transmission probability (b), and electron density (c) of the triple-well RTD self-consistently calculated at $T = 220$ K near equilibrium.

The electric field in W1 is largely screened by these accumulated electrons and the bottom of well W1 is kept virtually flat until the resonances occur at high applied voltages. The widths of the narrower wells, L_{W2} and L_{W3}, are chosen so that the condition $E_R^1 < E_F^L < E_R^2 < E_R^3$ can be satisfied. The transmission probability through these wells is very small in equilibrium, as can be seen in Fig. 5.23(a), but approaches 100% when the states Ψ_R^2 or Ψ_R^3 resonate with the injection level Ψ_R^1 under larger applied biases. Therefore, the current peak widths in the *I–V* characteristics are very small, and the valley currents are expected to be greatly decreased.

The *I–V* characteristics have been calculated using self-consistently obtained transmission probability. Figure 5.24 shows the *I–V* curve calculated at 220 K, at which temperature current peaks of nearly equal size are observed experimentally, as shown later in Fig. 5.28(a). The characteristics with double NDC can have three stable points with an appropriate load resistance since the valley current at the second off-resonance is smaller than the first peak current. The first and second current peaks result from resonances of Ψ_R^2 and Ψ_R^3 with Ψ_R^1. Energy-band diagrams, transmission probabilities and electron density distributions are shown in Fig. 5.25(a)–(c) at the three different states indicated by arrows in Fig. 5.24: at the first resonance (Fig. 5.25(a)), at the off-resonance (Fig. 5.25(b)) and at the second resonance (Fig. 5.25(c)).

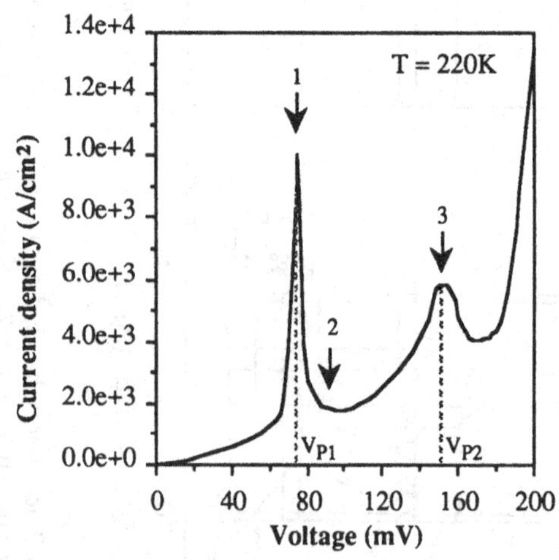

Figure 5.24 Calculated *I–V* characteristics of a triple-well RTD self-consistently calculated at 220 K. Three typical states are pointed out by the arrows: arrow 1 for the first current peak, 2 for the current valley and 3 for the second current peak.

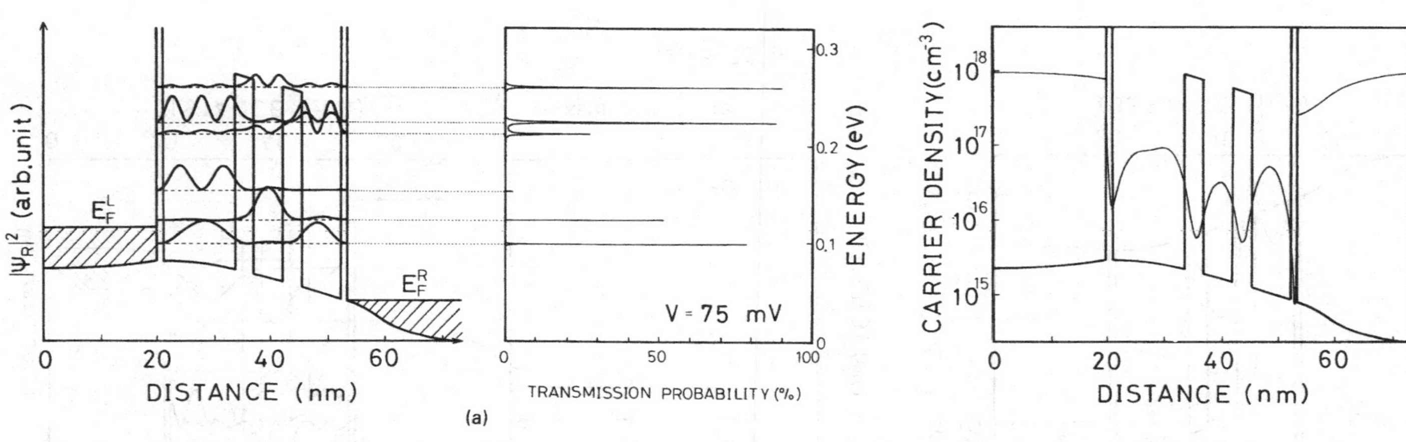

Figure 5.25 The energy-band diagram (left), transmission probability (center) and electron density (right) of a triple-well RTD self-consistently calculated at (a) the first resonance, and overleaf (b) the first resonance and (c) the second resonance.

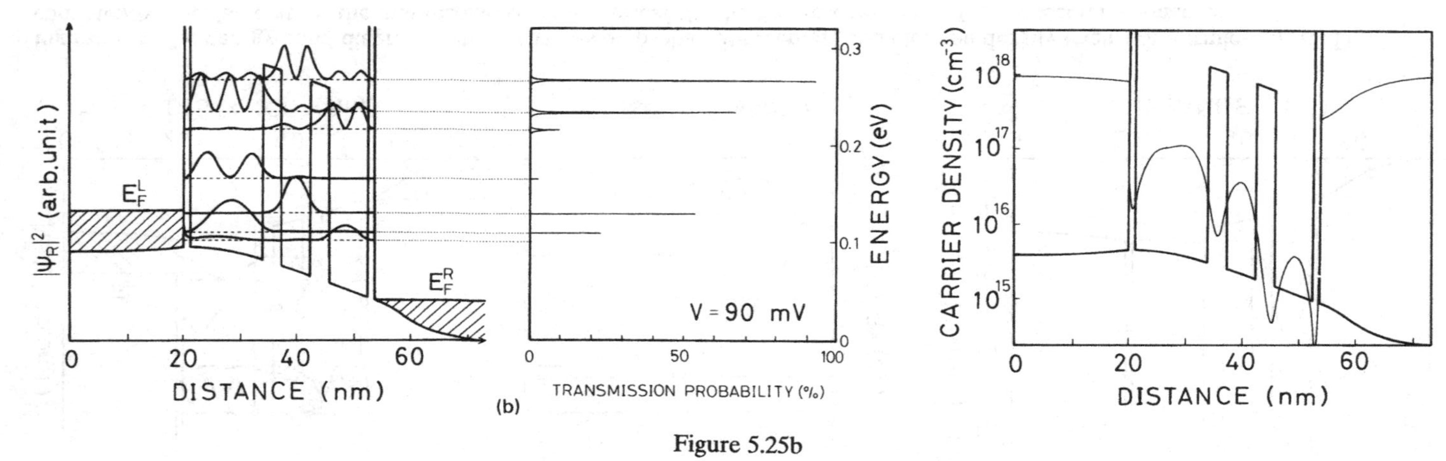
$|\Psi_R|^2$ (arb.unit)
E_F^L
E_F^R
DISTANCE (nm)
(b)
V = 90 mV
TRANSMISSION PROBABILITY (%)
ENERGY (eV)
CARRIER DENSITY (cm^{-3})
DISTANCE (nm)

Figure 5.25b

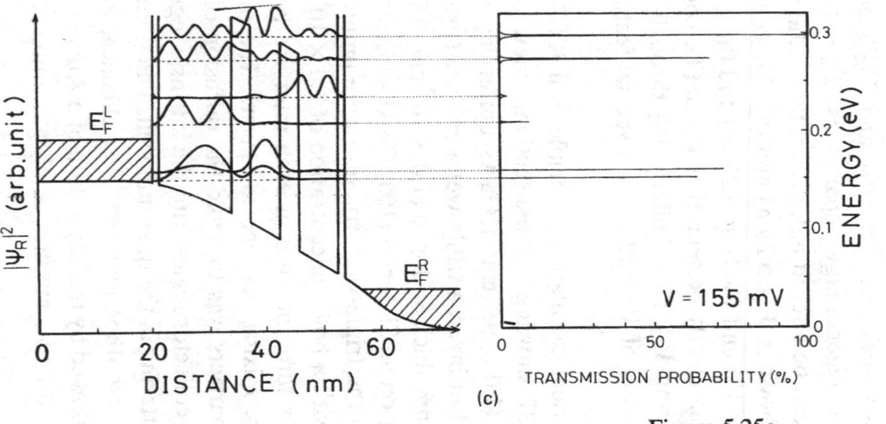

$|\Psi_R|^2$ (arb.unit)
E_F^L
E_F^R
0
20
40
60
DISTANCE (nm)
0.3
0.2
0.1
0
ENERGY (eV)
V = 155 mV
0
50
100
TRANSMISSION PROBABILITY (%)
(c)

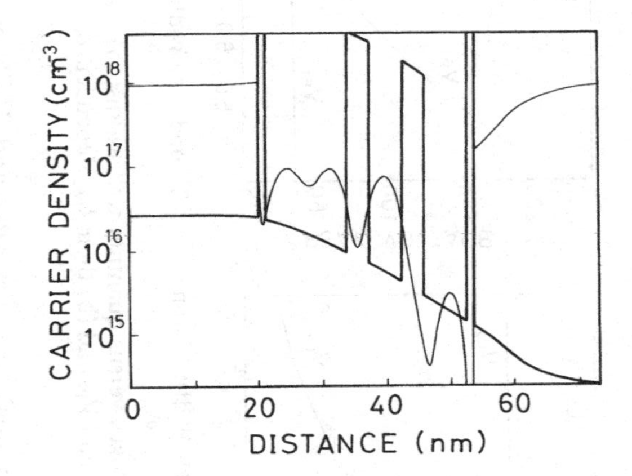

CARRIER DENSITY (cm^{-3})
10^{18}
10^{17}
10^{16}
10^{15}
0
20
40
60
DISTANCE (nm)

Figure 5.25c

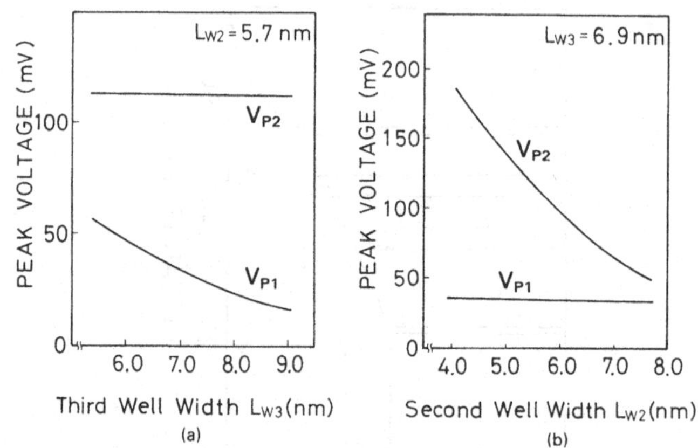

Figure 5.26 Peak voltages versus quantum well widths: (a) L_{W3} dependence of the peak voltages V_{P1} and V_{P2} and (b) their L_{W2} dependence.

Therefore, the two peak voltages, V_{P1} and V_{P2} (see Fig. 5.24), are determined by the eigenenergy separations, $E_R^2 - E_R^1$ and $E_R^3 - E_R^1$, respectively, and can be altered independently by the well thicknesses, L_{W3} and L_{W2}, as shown in Fig. 5.26(a) and (b). This diagram shows the L_{W3} dependence of V_{P1} and V_{P2} (Fig. 5.26(a)) and their L_{W2} dependence (Fig. 5.26(b)). It can clearly be seen that V_{P1} can be controlled through L_{W3} without changing V_{P2}, and similarly V_{P2} through L_{W2}. Such independent controllability of V_{P1} and V_{P2} is a special feature of this triple-well RTD.

Based on numerical calculations, a triple-well RTD has been fabricated [31], Fig. 5.27 showing a cross-sectional view of the device. A 500-nm Si-doped (1.0×10^{18} cm^{-3}) GaAs buffer layer, a 10-nm-thick undoped GaAs offset layer, a triple-well structure (listed in Table 5.2) and, finally, a 300-nm-thick Si-doped (1.0×10^{18} cm^{-3}) GaAs layer were grown successively on an Si-doped (100) GaAs substrate.

All the layers in the triple-well structure were unintentionally doped (typically p-type with a hole concentration of 1.0×10^{14} cm^{-3}). Growth was interrupted for three minutes at each interface between different materials to reduce mixing between the layers. The offset layer introduced under the barriers was to avoid the diffusion of ionised Si ions from the n^+-GaAs collector region into the intrinsic region which would cause impurity scattering in the quantum well. The wafer was processed by conventional photolithography and wet chemical etching to define the device area, followed by the deposition of a AuGe alloy, lift-off, and sintering at 400 °C for two minutes in N_2 gas ambient to form ohmic

Table 5.3. Calculated and observed peak voltages V_{P1} and V_P

	Calculated	Observed
First peak voltage V_{P1}(mV)	74	60
Second peak voltage V_{P2}(mV)	152	160

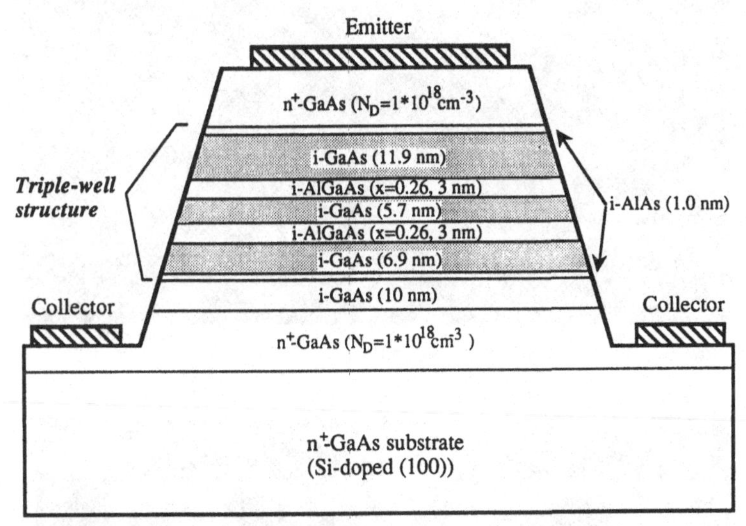

Figure 5.27 Schematic cross-sectional view of a fabricated triple-well RTD. Shaded regions represent three quantum wells.

contacts. Devices were fabricated with various sizes of top contact pads ranging from $4 \times 4\ \mu m^2$ to $100 \times 100\ \mu m^2$.

The *I–V* characteristics measured at 220 K and 300 K are shown in Fig. 5.28(a) and (b). The characteristics exhibit two current peaks with nearly the same amplitude at 220 K. It can be seen that the device exhibits a significant double NDC and the characteristics at 220 K can be used for triple-value logic applications.

The peak current dependencies of two peak voltages extracted from devices with various areas are shown in Fig. 5.29: both are found to depend almost linearly on the current through a resistance of approximately 4 Ω which is assumed to be due to the external measurement system in series with the device. Intrinsic values of the peak voltages are extracted from the values at the intersections of V_{P1} and V_{P2} with the vertical axis ($I_P = 0$). These intrinsic peak voltages are listed in Table 5.3. Both measured peak voltages are in a good agreement with those

(a)

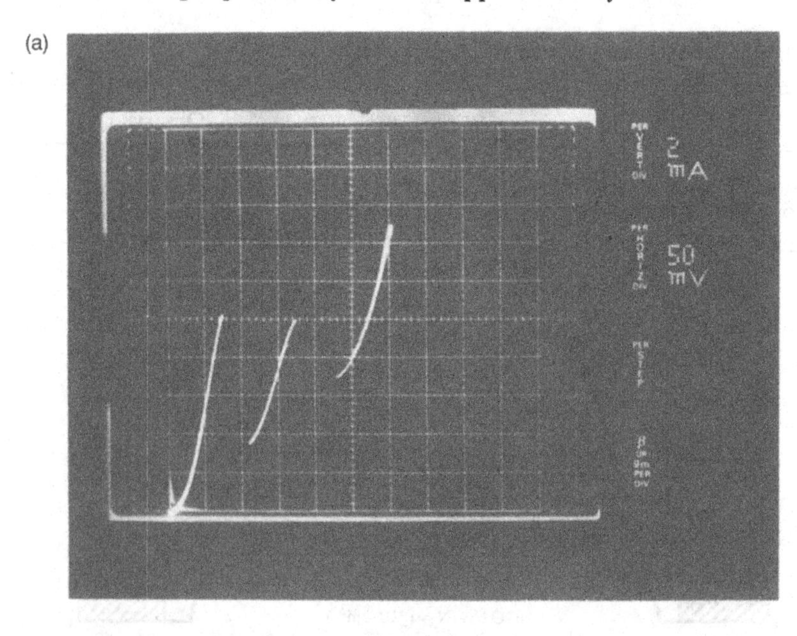

(b)

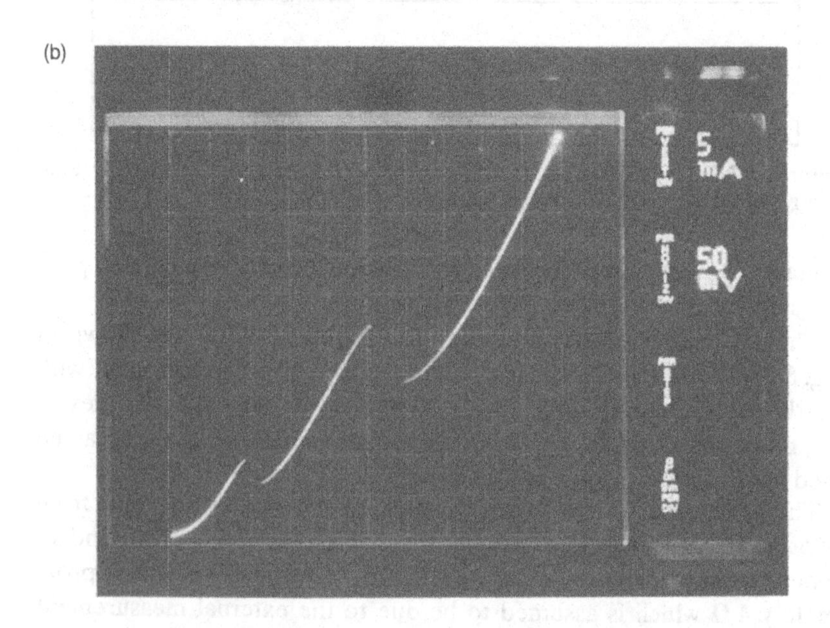

Figure 5.28 *I–V* characteristics of the triple-well RTD exhibiting significant double NDC measured at (a) 220 K and (b) room temperature. Data (a) were taken from a sample with a mesa area of 100 μm × 100 μm and (b) from that of 75 μm × 100 μm.

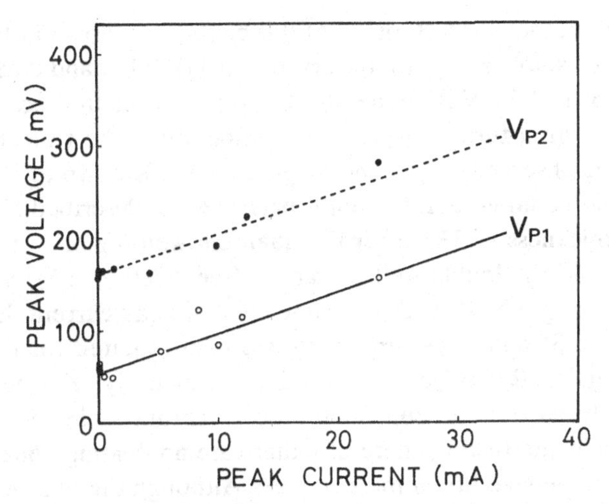

Figure 5.29 Experimental dependence of the two peak voltages on peak currents taken from devices with various sizes.

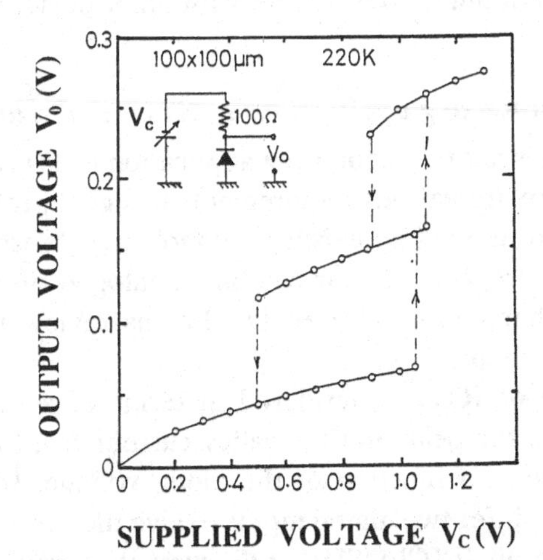

Figure 5.30 Input–output characteristics of a simple circuit which consists of a triple-well RTD and a load resistance of 100 Ω (see inset), measured at 220 K. Three stable points are seen at 0.054, 0.155 and 0.25 V.

obtained from the calculations, despite a large voltage decrease in the spacer region. This fact indicates that the present self-consistent modelling reproduces the overall band diagram quite well over the whole range of applied bias.

A simple circuit was designed using a device with an area of $100 \times 100\ \mu m^2$ with a load resistance of 100 Ω (see inset of Fig. 5.30).

The input–output characteristics of this circuit are shown in Fig. 5.30 in which three stable states are observed at 0.065, 0.155 and 0.250 V when an input bias of 1.0 V is applied to the circuit. Since the second current decreases with a further decrease in temperature, the tri-stable operation can be observed only in the range from 230 K down to 180 K.

It should be noted that the multi-barrier RTD described above has a layer of thickness of 33 nm for the quantum well region. Similarly, the InGaAs/InAlAs double-well structure (see Section 2.5), which also exhibits multiple NDC with the same level of peak current, has a layer thickness of 35 nm. These are much less than required for the stacked double-barrier RT structures described previously. Furthermore, in these multi-barrier structures, all the quantum wells are quantum mechanically interactive; therefore, there are no floating nodes to cause potential multi-stability in the structure. Although the number of NDC peaks realised by these multi-barrier structures is smaller than that realised by stacked double barriers, this approach is promising for the future integration of multi-NDC diodes with other devices.

5.2.3 Signal processing applications of RTDs and related devices

Signal processing is another promising application of RTDs and related devices since there are various requirements in signal processing which have not yet been fulfilled by existing technologies. Most of the signal processing circuits reported so far combine analogue signal processing with the NDC characteristics of RTDs. Let us review the operating principle of these circuits.

The *I–V* curve of RTDs, considered in terms of a certain current threshold between the peak and the valley current, is a low–high–low–high characteristic as a function of the diode voltage. If one uses an analogue sum circuit for two digital inputs to give the supply voltage, it is possible to create an EXCLUSIVE OR current output, in which the current becomes high when only one of the inputs is high; the output can be set to become low due to the NDC characteristics for the two high-input signal by choosing proper summation characteristics at the input, as shown in Fig. 5.31. This characteristic can be utilised to create a parity-bit generator circuit, which is a circuit that determines whether the total number of high inputs among all the inputs is even or odd. In other words, it is a digital adder circuit without the carry signal. The truth table for an adder circuit is the same as that of an EXCLUSIVE OR circuit, as shown in Fig. 5.31. If one expands this scheme to the

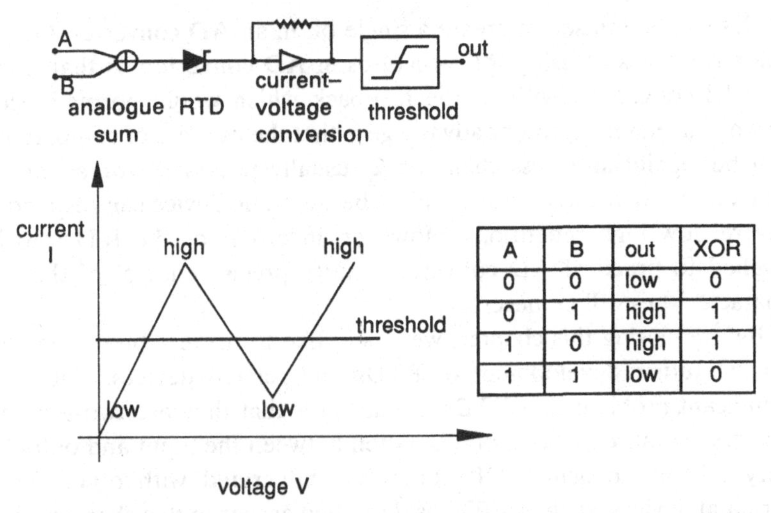

A	B	out	XOR
0	0	low	0
0	1	high	1
1	0	high	1
1	1	low	0

Figure 5.31 An EXCLUSIVE OR circuit using an RTD with low–high–low–high characteristic and the truth table for the circuit.

multiple NDC diodes described in the preceding section, one obtains repeated low–high characteristics. A natural continuation of the above discussion results in a multi-bit parity-bit generator utilising a multiple-NDC diode. By using five stacks of double-barrier RTDs, 11-bit parity generation is reported to be possible [33]. In another report [34], a vertically integrated device consisting of two double-barrier structures and a bipolar transistor has been applied to a four-bit parity generator. In this case, the bipolar transistor is connected to external resistors to form the input analogue sum circuit. In conventional circuit technology, a parity-bit generator needs as many EXCLUSIVE OR gates as the number of bits minus one, which are combined in several stages; this means that the data processing requires several steps and is thus time consuming. In contrast, the circuit using RT structures has a smaller number of devices and only one stage. Therefore, it will be able to reduce both the area of the circuit and the time required to create the parity-bit signal.

Another possibility for utilising the low–high–low–high current characteristics is in the analogue-to-digital (AD) converter. In an AD converter, the least significant bit changes between 0 and 1 repeatedly as the input changes monotonically, the next significant bit changing its value for each repetition. Therefore, if used with proper voltage scaling networks and biasing circuits, the low–high repetition characteristics of

RTDs can be utilised to create a single bit in an AD converter. One of the advantages of using RT structures in AD converters is that it is a parallel conversion without the feedback which would normally slow down conversion. Another advantage is that the number of comparators can be significantly less than the 2^n usually necessary for a parallel converter with n-bit accuracy. This is because one device can respond to several low–high repetitions. However, in order for the RTDs to be applied to practical AD converters, more precise control of the I–V characteristics will be necessary.

Before closing this chapter, we would like to make a brief comment on the future development of RTDs and related devices. The most significant problem that RTDs are facing is that they are two-terminal devices. Because of this, good isolation between the input and output is very difficult to achieve. If RTDs are integrated with other three-terminal devices, such as FETs as described earlier in this chapter, good isolation is readily obtained; however, speed is limited in the three-terminal device since any three-terminal device has parasitic capacitance and resistance related to the control terminal, such as the gate of an FET or the base of a bipolar transistor. To eliminate this speed-limiting factor, the only possible way seems to be to use optical signals to control the device. Although no device has yet appeared in which the resonant tunnelling current is directly controlled by an optical signal, it is expected that this kind of device could handle high-frequency signals in the range of terahertz frequencies since optical pulses shorter than 0.1 ps are becoming available in laboratory environments. With regard to functional applications, it seems that the integration of RTDs with existing three-terminal devices has produced at most an order of magnitude improvement in functionality such as the reduction of the number of devices, power consumption or improvement in the speed of operation. In order to realise new devices with more than an order of magnitude improvement in performance over conventional devices, ways other than simple integration of RTDs with conventional devices should be pursued. Consequently, much attention should be paid to studies of resonant tunnelling transistors [35], [43]. Only by effective, functional utilisation of resonant tunnelling phenomenon through existing technologies will the 'functional' device be realised.

5.3 References

[1] E. R. Brown, T. C. L. G. Sollner, C. D. Parker, W. D. Goodhue, and C. L. Chen, Oscillations up to 420 GHz in GaAs/AlAs resonant tunneling diodes, *Appl. Phys. Lett.*, **55**, 1777, 1989.

[2] E. R. Brown, J. R. Söderström, C. D. Parker, L. J. Mahoney, K. M. Molvar and T. C. McGill, Oscillations up to 712 GHz in InAs/AlSb resonant tunneling diodes, *Appl. Phys. Lett.*, **58**, 2291, 1991.

[3] T. C. L. G. Sollner, E. R. Brown, C. D. Parker and W. D. Goodhue, High-frequency applications of resonant-tunneling diodes, *Electronic Properties of Multilayers and Low-dimensional Semiconductor Structures*, edited by J. M. Chamberlain *et al.* (Plenum Press, New York, 1990), pp. 283–96.

[4] E. R. Brown, Resonant tunneling in high-speed double barrier diodes, v. 3 in *Hot Carriers in Semiconductor Nanostructures: Physics and applications*, American Telephone & Telegraph, 1991.

[5] E. R. Brown, C. D. Parker and T. C. L. G. Sollner, Effect of quasi-bound state lifetime on the oscillation power of resonant tunneling diodes, *Appl. Phys. Lett.*, **54**, 934, 1989.

[6] R. K. Mains and G. I. Haddad, Time-dependent modeling of resonant tunneling diodes from direct solution of the Schrödinger equation, *J. Appl. Phys.*, **64**, 3564, 1988.

[7] W. R. Frensley, Wigner-function model of a resonant tunneling semiconductor device, *Phys. Rev.*, **B36**, 1570, 1987.

[8] T. P. E. Broekaert and C. G. Fonstad, $In_{0.53}Ga_{0.47}As/AlAs$ resonant tunneling diodes with peak current densities in excess of 450 kA/cm^2, *J. Appl. Phys.*, **68**, 4310, 1990.

[9] J. R. Söderström, E. R. Brown, C. D. Parker, L. J. Mahoney, J. Y. Yao, T. G. Andersson and T. C. McGill, Growth and characterization of high current density, high speed InAs/AlSb resonant tunneling diodes, *Appl. Phys. Lett.*, **58**, 275, 1991.

[10] V. P. Kesan, D. P. Neikirk, B. G. Streetman and P. A. Blakey, The influence of transit-time effects on the optimum design and maximum oscillation frequency of quantum well oscillators, *IEEE Trans. Electron Devices*, **ED-35**, 405, 1988.

[11] T. Nittono, H. Ito, O. Nakajima and T. Ishibashi, Non-alloyed ohmic contacts n-GaAs using compositionary graded $In_xGa_{1-x}As$ layers, *Japan. J. Appl. Phys.*, **27**, 1718, 1988.

[12] A. V. Dyadchenko and E. D. Prokhorov, Electron drift velocity in $In_xGa_{1-x}As$ compounds, *Radio Eng. and Electron Phys.*, **21**, 151, 1976.

[13] T. Ishibashi and Y. Yamauchi, A possible near-ballistic collection in an AlGaAs/GaAs HBT with a modified collector structure, *IEEE Trans. Electron Devices*, **ED-35**, 401, 1988.

[14] P. D. Batelaan and M. A. Frerking, Conference Digest, 12th International Conference on infrared and millimeter waves, edited by R. J. Temkin (IEEE, New York), p. 14.

[15] T. C. L. G. Sollner, E. R. Brown, W. D. Goodhue and C. A. Correa, Harmonic multiplication using resonant tunneling, *J. Appl. Phys.*, **64**, 4248, 1988.

[16] A. Rydberg and H. Grönqvist, Quantum-well high-efficiency millimetre-wave frequency tripler, *Electron. Lett.*, **25**, 1989.

[17] J. F. Whitaker, G. A. Mourou, T. C. L. G. Sollner and W. D. Goodhue,

Picosecond switching time measurement of a resonant tunneling diode, *Appl. Phys. Lett.*, **53**, 385, 1988.
[18] S. K. Diamond, E. Özbay, M. J. W. Rodwell, D. M. Bloom, Y. C. Pao and J. S. Harris, Resonant tunneling diodes for switching applications, *Appl. Phys. Lett.*, **54**, 153, 1989.
[19] E. Özbay, D. M. Bloom, D. H. Chow and J. N. Schulman, 1.7 ps, microwave, integrated circuit compatible InAs/AlSb resonant tunneling diodes, *IEEE Electron Device Lett.*, **EDL-14**, 400, 1993.
[20] E. Özbay, D. M. Bloom and S. K. Diamond, Pulse forming and triggering using resonant tunneling diode structures, *Electron. Lett.*, **26**, 1046, 1990.
[21] E. Özbay and D. M. Bloom, 110-GHz monolithic resonant-tunneling-diode trigger circuit, *IEEE Electron Device Lett.*, **EDL-12**, 480, 1991.
[22] E. R. Brown, C. D. Parker, A. R. Calawa, M. J. Manfra, T. C. L. G. Sollner, C. L. Chen, S. W. Pang and K. M. Molvar, High-speed resonant tunneling diodes made from $In_{0.53}Ga_{0.47}As/AlAs$ material system, *Proc. SPIE*, **1288**, 122, 1990.
[23] K. L. Lear, K. Yoh and J. S. Harris, Jr, Monolithic integration of GaAs/AlGaAs resonant tunnel diode load and GaAs enhancement-mode MESFET drivers for tunnel diode FET logic gates, *Proc. Int. Symp. GaAs and Related Compounds, Atlanta, Georgia, 1988*, IOP Publishing Ltd, p. 593, 1988.
[24] E. R. Brown, M. A. Hollis, F. W. Smith, K. C. Wang and P. M. Asbeck, Resonant-tunneling-diode loads: Speed limits and application in fast logic circuits, in *ISSCC Dig. Tech. Papers*, pp. 142–3, 1992.
[25] Y. Watanabe, Y. Nakasha, K. Imanishi and M. Takikawa, Monolithic integration of InGaAs/InAlAs resonant tunneling diodes and HEMT for single transistor cell SRAM application, *Tech Dig.*, IEDM 92, pp. 475–8, 1992.
[26] T. Mori, S. Mutoh, H. Tamura and N. Yokoyama, An SRAM cell using double emitter RHET for gigabit-plus memory applications, *Extended Abstracts of 1993 Int. Conf. on Solid State Devices and Materials, Makuhari*, The Japan Society of Applied Physics, pp. 1074–6, 1993.
[27] F. Capasso, S. Sen, A. Y. Cho and D. Sivco, Resonant tunneling devices with multiple negative differential resistance and demonstration of a three-state memory cell for multiple-valued logic applications, *IEEE Electron Device Lett.*, **EDL-8**, 297, 1987.
[28] F. Capasso and R. A. Kiehl, Resonant tunneling transistor with quantum well base and high-energy injection: A new negative differential device, *J. Appl. Phys.*, **58**, 1366, 1985.
[29] A. A. Lakhani and R. C. Potter, Combining resonant tunneling diodes for signal processing applications, *Appl. Phys. Lett.*, **52**, 1684, 1988.
[30] J. R. Söderström and T. G. Andersson, A multiple-state memory cell based on the resonant tunneling diode, *IEEE Electron Device Lett.*, **EDL-9**, 200, 1988.
[31] T. Tanoue, H. Mizuta and S. Takahashi, A triple-well resonant tunneling diode for multiple-valued logic application, *IEEE Electron Device Lett.*, **EDL-9**, 365, 1988.
[32] H. Mizuta, T. Tanoue and S. Takahashi, A new triple-well resonant tunneling diode with controllable double negative resistance, *IEEE Trans. Electron Devices*, **ED-35**, 1951, 1988.
[33] A. A. Lakhani, R. C. Potter and H. S. Hier, Eleven-bit parity bit

generator with a single, vertically integrated resonant tunneling device, *Electron. Lett.*, **24**, 681, 1988.
[34] S. Sen, F. Capasso, A. Y. Cho and D. Sivco, Parity generator circuit using a multi-state resonant tunneling bipolar transistor, *Electron. Lett.*, **24**, 1506, 1988.
[35] Various circuits using RHETs are reviewed in: N. Yokoyama, H. Ohnishi, T. Mori, M. Takatsu, S. Mutoh, K. Imamura and A. Shibatomi, Resonant tunneling hot electron transistor, in v. 2 in *Hot Carriers in Semiconductor Nanostructures: Physics and applications*, American Telephone & Telegraph, 1991.
[36] T. Akeyoshi, K. Maezawa and T. Mizutani, Weighted sum logic operation in multiple input MOBILEs (Monostable-Bistable Transition Logic Elements), *Extended Abstracts of 1993 Int. Conf. Solid State Devices and Materials, Makuhari*, The Japan Society of Applied Physics, pp. 1077–9, 1993.
[37] K. Lehovec, GaAs enhancement mode FET-tunnel diode ultra-fast low power inverter and memory cell, *IEEE J. Solid-State Circuits*, **14**, 797–800, 1979.
[38] R. H. Bergman, M. Cooperm and H. Ur, High speed logic circuits using tunnel diodes, *RCA Review*, 152–286, 1962.
[39] A. R. Bonnefoi, T. C. McGill and R. D. Burnham, Resonant tunneling transistors with controllable negative differential resistances, *IEEE Electron Device Lett.*, **EDL-6**, 636, 1985.
[40] T. K. Woodward, T. C. McGill, H. F. Chung and R. D. Burnham, Integration of a resonant tunnelling structure with a metal-semiconductor field effect transistor, *Appl. Phys. Lett.*, **51**, 1542, 1987.
[41] K. K. Gullapalli, A. J. Tsao and D. P. Neikirk, Observation of zero-bias multi-state behavior in selectively doped two-terminal quantum tunneling devices, in *Tech. Dig.*, IEDM'92, 479, 1992.
[42] C. J. Goodings, Variable-area resonant tunnelling diodes using implanted gates, PhD thesis, Cambridge University, 1993.
[43] A. C. Seabaugh, Y. C. Kao, W. R. Frensley, J. N. Randall and M. A. Reed, Resonant transmission in the base/collector junction of a bipolar quantum-well resonant-tunneling transistor, *Appl. Phys. Lett.*, **59**, 3413, 1991.

6

Resonant tunnelling in low-dimensional double-barrier heterostructures

This last chapter is devoted to the study of resonant tunnelling through laterally confined, ultra-small, double-barrier heterostructures [1]–[8]. Recent rapid advances in nanofabrication techniques have naturally led to the idea of resonant tunnelling through three-dimensionally confined 'quantum dot' structures. Since electrons are confined laterally as well as vertically in these structures, the devices are often called *zero-dimensional (0D) RTDs* and have become of great interest both from the standpoint of the physics of quantum transport through 0D electronic states and also for device miniaturisation towards highly integrated functional resonant tunnelling devices. The 0D RTD is a virtually isolated quantum dot only weakly coupled to its reservoirs and thus is well suited to investigating electron-wave transport properties through 3D quantised energy levels. By designing the structural parameters such as the barrier thickness, the quantum well width and dimensionality of lateral confinement, it is possible to realise a 'quantum dot' in which the number of electrons is nearly quantised so the effect of single-charge-assisted transport, or the so-called Coulomb blockade (CB) [9]–[11], becomes significant. After Reed *et al.* reported their pioneering work in 1988 on resonant tunnelling through a quantum pillar which was fabricated by electron beam lithography and dry etching, several theoretical and experimental studies have been reported which investigate the mechanism of the observed fine structures. Transport in the 0D RTD is generally much more complicated than that in the conventional large-area RTDs which we have studied so far in this book: problems such as lateral-mode mixing due to a non-uniform confinement potential, charge quantisation in a quantum well and the interplay between resonant tunnelling and Coulomb blockade single-electron tunnelling have recently been invoked for the 0D RTDs. Such difficulties are still far from being fully resolved.

The various 0D RTDs reported so far are reviewed in Section 6.1, and the basic idea of lateral quantisation is studied. Section 6.2 presents a theoretical study of resonant tunnelling through 0D RTDs which is based on 3D scattering matrix (*S-matrix*) theory (Section 6.2.1): the theoretical modelling introduced in this chapter is a natural expansion of the global coherent tunnelling model that we studied in Chapter 2 into 3D systems. We use this model to study lateral-mode non-conserving resonant tunnelling which results from non-uniform lateral confinement (Section 6.2.2). Experimental studies of 0D RTDs are presented in Section 6.3 using a novel, variable-area resonant tunnelling diode (VARTD) in which the size of the lateral confinement can be altered (Section 6.3.1). Fine structures observed in the *I–V* characteristics of these VARTDs are discussed in terms of the effects of 3D electronic confinement (Section 6.3.2) and single-impurity-assisted tunnelling (Section 6.3.3). We also study 1D RTDs in which the electrons are laterally confined in one direction only. Finally, the interplay between the RT and Coulomb blockade expected in 0D RTDs is discussed briefly in Section 6.5.

6.1 Low-dimensional resonant tunnelling structures

The pioneering work on resonant tunnelling through a 3D confined quantum well, a quantum dot, was performed by Reed *et al.* [1], [2]. The conventional resonant tunnelling wafer was etched vertically down to the n^+-GaAs bottom contact layer so that quantum pillars (see Fig. 6.1) were defined on the wafer. Reed *et al.* fabricated a collection of quantum pillars with diameters in the range 100–250 nm. The initial epitaxial structure was a 5 nm undoped $In_{0.08}Ga_{0.92}As$ strained well sandwiched between 4 nm undoped $Al_{0.25}Ga_{0.75}As$ barriers with 10-nm-thick undoped GaAs spacer layers and 20-nm-thick n-GaAs layers with a graded doping profile grown on an n^+-GaAs contact layer. Electron beam lithography and highly anisotropic reactive ion etching were used to fabricate these structures, with the ohmic contact metal on top. A similar structure has also been used by Su *et al.* [4] using wet chemical etching. In these etched structures there exist defect states on the lateral surface of the pillar which capture electrons, and the density of the surface states is sufficiently high to pin the Fermi-level at the surface. Thus electrons in the nanostructure are laterally confined in the gutter-

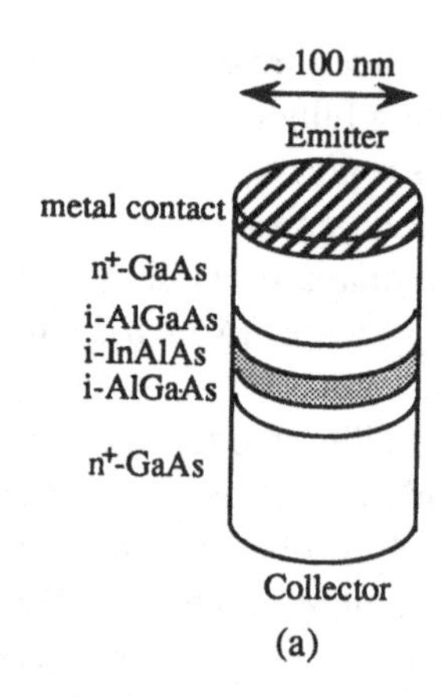

(a)

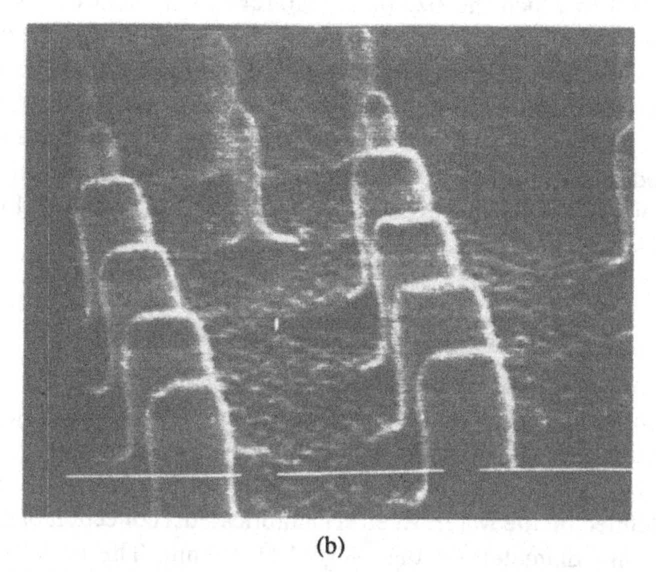

(b)

Figure 6.1 (a) A schematic diagram of the simply etched quantum pillars fabricated by Reed *et al.* (b) Scanning electron micrograph of the fabricated pillars. After Reed *et al.* [1], with permission.

shaped potential well, shown in Fig. 6.2, which consists of a parabolic part ($r_c < r < r_s$) and a flat part ($0 < r < r_c$). This sort of nanometre-scale device is hereafter called a *zero-dimensional resonant tunnelling diode* (0D RTD) as compared with conventional large-area RTDs in which the motion of the electrons is not quantised in a 2D plane parallel to the heterointerface, and should thus be referred to as 2D RTDs in this chapter.

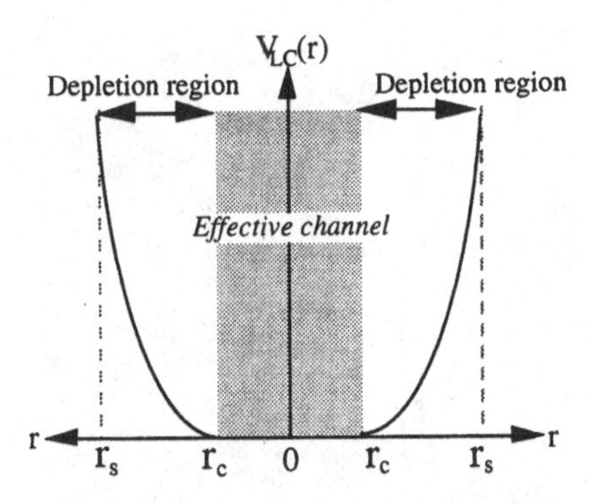

Figure 6.2 A gutter-shaped lateral confinement potential formed in the simply etched structure.

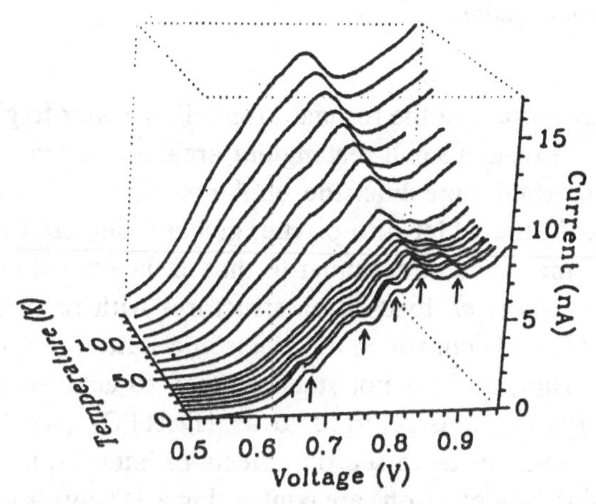

Figure 6.3 Current–voltage–temperature characteristics of a single-quantum-dot nanostructure with a diameter of 100 nm. The arrows indicate the voltage peak positions of discrete state tunnelling for the $T = 1.0$ K curve. After Reed *et al.* [1], with permission.

Reed *et al.* measured the *I–V* characteristics of the single pillars at various temperatures [1], [2]. As shown in Fig. 6.3 at high temperatures the 0D RTD shows an NDC in the *I–V* characteristics which is basically the same as that of the 2D RTDs. On lowering the temperature, however, a series of small current peaks appear which are superimposed on the NDC. These fine structures were attributed to resonant tunnelling through 3D quantised states in the quantum dot.

A different type of 0D RTD has been reported by Tarucha *et al.* [3], [5]. Instead of etching a pillar structure, they used focused Ga-ion beam implantation to define a 0D structure. A 100-nm-diameter Ga-FIB was

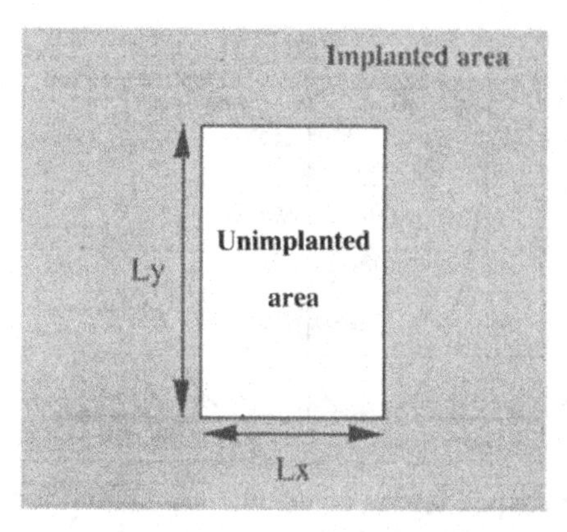

Figure 6.4 Schematic top view of a 0D structure fabricated by Tarucha *et al.* by Ga ion-beam implantation.

scanned on the surface of the resonant tunnelling wafer to give a dose of 5×10^{12} cm^{-2}, leaving a small rectangular area, as illustrated in Fig. 6.4. After rapid thermal annealing, the implanted Ga ions convert n-type GaAs regions in the wafer into p-type, giving a depleted region and a lateral confinement potential similar to that of the etched nanostructure (Fig. 6.2). Tarucha *et al.* focused on structures with rectangular lateral confinement ($L_x < L_y$) in which the electronic states are quantised only in the x-dimension as L_y is not small enough to achieve quantisation: these structures are thus referred to as 1D RTDs (see Section 6.4). This structure enables us to see the effects of lateral quantisation in a clearer way since the electrons are confined in a 1D gutter potential well resulting in nearly equal splitting of the quantised energy, E_x, in the x-dimension. Tarucha *et al.* measured the I–V characteristics for the 1D RTDs with various areas ($L_x \times L_y$) at 4 K [3]. As shown in Fig. 6.5, the largest sample, with an L_x of 310 nm, shows only a single NDC associated with the first quasi-bound state of the double-barrier structure. A series of small current peaks is observed for the smaller samples in the NDC regime which are attributable to the quantisation in the x-dimension and these are more pronounced, with a layer separation in the case of the smaller device. Tarucha *et al.* analysed the observed peak voltages by comparing them with the subband energies calculated for the 1D gutter potential and assigned these peaks up to the seventh quantised level [3].

A theoretical description of the 0D RT is much more complicated than that for the conventional 2D RT because of lateral confinement.

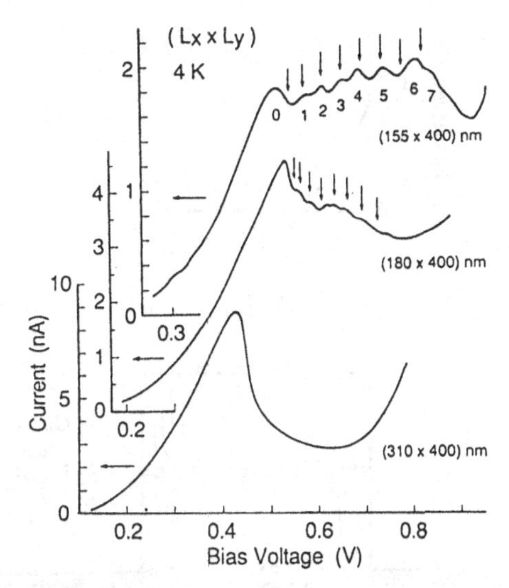

Figure 6.5 *I–V* characteristics of diodes with various unimplanted areas ($L_x \times L_y$). The arrows indicate the calculated voltage positions of the discrete states to be compared with the fine structures. After Tarucha *et al.* [3], with permission.

The difficulty stems from the *non-uniform* lateral confinement potential which is formed by variation in the surface depletion, even though the device is fabricated to be geometrically straight in the vertical direction.

Figure 6.6 shows an example of lateral confinement potential distribution which resembles an hourglass along the channel: the detail of this diagram is explained in Section 6.2.2. This non-uniform confinement mainly results in the following two effects. Firstly, the lateral quantised energies vary steeply in the z-dimension around the quantum dot. The energy separation between the lateral energies in the quantum well is larger than that in the emitter and collector regions. Secondly, the non-uniform confinement causes lateral wavefunction mismatch and mixes the lateral electronic states (*lateral-mode mixing*): theoretical work reported by Bryant [12]–[14] has demonstrated the importance of lateral-mode mixing in this system. It is therefore difficult to construct an intuitive picture of the 0D RT, and proper theoretical modelling requires 3D transport theory, as shown in Section 6.2.1. A rather simple model, however, was proposed by Reed *et al.* [2] based on transfer Hamiltonian modelling (Section 2.2.3) and it is worth introducing this before proceeding to the 3D theory.

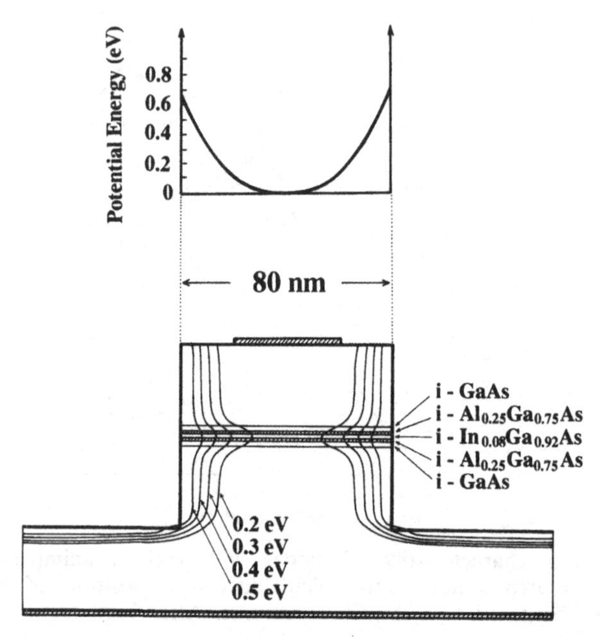

Figure 6.6 Lateral confinement potential distribution in a simply etched 0D resonant tunnelling structure calculated numerically using a classical hydrodynamic device simulation taking account of surface carrier trap levels.

Following the early theoretical work by Bryant [13], Reed *et al.* introduced two quantum numbers γ and γ' to describe the lateral motion of electrons in the emitter region and in the quantum well respectively. Thus the energies of electrons in these two regions are expressed approximately as $\epsilon_\gamma + E_{k_z}$ and $\epsilon_\gamma' + E_{k_{z0}}$, where ϵ_γ is the energy of the γth lateral mode, E_{k_z} is the z-component of energy $(= \hbar^2 k_z^2/2m^*)$ and $E_{k_{z0}}$ is the energy of the lowest quasi-bound state in the quantum well. The interval between the lateral energy states $\epsilon_\gamma'+1 - \epsilon_\gamma'$ in the quantum well is larger than $\epsilon_{\gamma+1} - \epsilon_\gamma$ in the emitter region. A series of quantised levels in these regions is shown schematically in Fig. 6.7(a). Reed *et al.* reported that the observed fine structure can be modelled as resonances of the levels in the well with those in the emitter under an applied bias (see Fig. 6.7(b)). Each time a level in the well crosses one of the levels in the emitter, electrons tunnel through to the collector region, leading to a resonant current peak. If $\gamma = \gamma'$ the tunnelling conserves the lateral mode (lateral momentum) and if not it is lateral-mode non-conserving tunnelling. To evaluate transition rates and selection rules for these tunnelling processes, 3D numerical calcula-

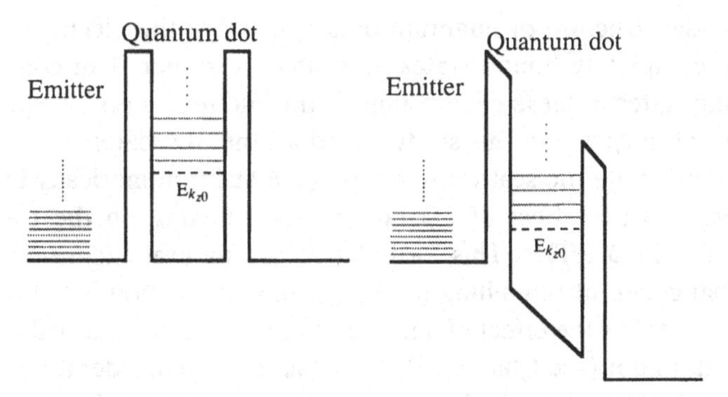

Figure 6.7 Schematic energy-band diagram with lateral eigenstates in the emitter and quantum dot (a) at zero bias and (b) under an applied bias.

tions using the profile of the confinement potential are apparently necessary. Reed *et al.* identified the crossings of the quantum well levels with the emitter levels as a function of applied bias, and compared them with the observed peak voltages [2]. The calculated results agree with the experimental results [2]: this fact indicates that lateral-mode non-conserving tunnelling actually contributes to the current. Tarucha *et al.* also adopted this model to analyse the experimental results of the 1D RTDs [5] and found that fine structure observed at low temperature is attributable to lateral-mode non-conserving tunnelling. This simple model gives us a good understanding of low-dimensional resonant tunnelling. In Section 6.2 further quantitative analysis of multi-mode resonant tunnelling is presented by introducing 3D S-matrix theory, which fully includes lateral-mode mixing.

6.2 Theory of zero-dimensional resonant tunnelling

6.2.1 Three-dimensional S-matrix theory

In this section a multi-mode *S*-matrix theory based on the 3D Schrödinger equation for open systems is presented in order to look into the detail of electron transport through 0D RTDs. Several theoretical studies of the 2D scattering equation have been reported for laterally patterned 2D electron gas systems [16]–[23]. Because direct numerical calculations usually involve a large amount of computational time and memory requirement, several useful alternative methods have also been proposed. The 3D Schrödinger equation has also been solved by Kumar for a completely isolated quantum dot under a magnetic field [24] and a

1D periodic structure of quantum dots [25]. Since the electronic states become completely bound states or subbands, rather than continuous scattering states in these circumstances, this method is not adequate for the present purpose. In this study we adopt the 3D scattering formulation and calculate the scattering matrix (S-matrix) numerically in order to investigate the effect of 2D lateral-mode mixing on the transport properties of 0D RTDs. This formulation is a natural expansion of the 1D global coherent tunnelling model (studied in Section 2.2.1) into 3D systems, and thus the effect of a non-equilibrium electron distribution in the quantum dot (see Chapter 4) is not taken into consideration. In the present calculation the self-consistent Hartree potential (see Section 2.2.2), which would be crucial to analysis of the Coulomb blockade, is neglected for simplicity.

We start from the 3D time-independent Schrödinger equation:

$$-\frac{\hbar^2}{2m^*}\left(\frac{\partial^2}{\partial x^2}+\frac{\partial^2}{\partial y^2}+\frac{\partial^2}{\partial z^2}\right)\Psi(x,y,z) + V(x,y,z)\Psi(x,y,z) = E\Psi(x,y,z) \tag{6.1}$$

where m^* is the conduction-band effective mass and $V(x,y,z)$ is the 3D potential distribution which consists of the electron affinity, $V_0(z)$ and the potential due to an external bias, $V_{EX}(z)$ (both assumed to be dependent on the z-coordinate only), the lateral confinement potential $V_{LC}(x,y,z)$, and any other scattering potentials such as the single ionised donor potential, $V_{IM}(x,y,z)$, discussed in Section 6.3.3:

$$V(x,y,z) = V_{LC}(x,y,z) + V_0(z) + V_{EX}(z) + V_{IM}(x,y,z) \tag{6.2}$$

The 3D wavefunction, $\Psi(x,y,z)$, is decomposed by using a complete set of 2D lateral wavefunctions at each z point, $\varphi_\gamma(x,y|z)$, as follows:

$$\Psi(x,y,z) = \sum_\gamma \varphi_\gamma(x,y|z)\chi_\gamma(z) \tag{6.3}$$

The z-dependent lateral wavefunction, $\varphi_\gamma(x,y|z)$ is obtained by solving the following 2D Schrödinger equation numerically:

$$-\frac{\hbar^2}{2m^*}\left(\frac{\partial^2}{\partial x^2}+\frac{\partial^2}{\partial y^2}\right)\varphi_\gamma(x,y|z) + V_{LC}(x,y,z)\varphi_\gamma(x,y|z) = \epsilon_\gamma(z)\varphi_\gamma(x,y|z) \tag{6.4}$$

with the Dirichlet boundary conditions, $\varphi_\gamma(x,y|z) = 0$, at the boundaries of the device. The index, γ, represents a 2D lateral-mode number and $\epsilon_\gamma(z)$ represents the corresponding γth lateral eigenenergy.

Substituting eqn (6.3) into eqn (6.1), the 3D Schrödinger equation reduces to the following 1D scattering equation for the z-component of the wavefunction $\chi_\gamma(z)$:

$$\frac{d^2}{dz^2}\chi_\gamma(z) + k_\gamma^2(z)\chi_\gamma(z) + \sum_{\gamma'}\left(2C_{\gamma,\gamma'}^{(0,1)}(z)\frac{d}{dz}\chi_{\gamma'}(z) + C_{\gamma,\gamma'}^{(0,2)}(z)\chi_{\gamma'}(z)\right) = \quad (6.5)$$

where $k_\gamma(z)$ denotes a complex wavenumber given by

$$k_\gamma^2(z) = \frac{2m^*}{\hbar^2}(E - \epsilon_\gamma(z) - V_0(z) - V_{EX}(z)) \quad (6.6)$$

Mode mixing coefficients, $C_{\gamma,\gamma'}^{(0,1)}$ and $C_{\gamma,\gamma'}^{(0,2)}$, are written as

$$C_{\gamma,\gamma'}^{(0,1)}(z) = \int dx \int dy \varphi_\gamma(x,y|z)\frac{\partial\varphi_{\gamma'}(x,y|z)}{\partial z} \quad (6.7)$$

$$C_{\gamma,\gamma'}^{(0,2)}(z) = \int dx \int dy \varphi_\gamma(x,y|z)\frac{\partial^2\varphi_{\gamma'}(x,y|z)}{\partial z^2} \quad (6.8)$$

and are evaluated by using the previously obtained set of lateral wavefunctions. The third term in eqn (6.5) causes mixing of lateral modes and is non-zero unless the system is completely uniform in the z-direction. The first derivative term of $\chi_\gamma(z)$ in eqn (6.5) can be eliminated by applying the unitary transformation:

$$\chi_\gamma(z) = \sum_{\gamma'} M_{\gamma,\gamma'}(z) f_{\gamma'}(z) \quad (6.9)$$

where the unitary matrix, $M_{\gamma\gamma'}(z)$, is defined as follows:

$$M_{\gamma,\gamma'}(z) = \exp\left(-\int^z C_{\gamma,\gamma'}^{(0,1)}(z')dz'\right) \quad (6.10)$$

The matrix $M_{\gamma\gamma'}(z)$ is calculated using a second-order expansion approximation [26] which guarantees the unitarity of the matrix. Substituting eqns (6.9) and (6.10) into eqn (6.5), the transformation leads to the following equation:

$$\frac{d^2}{dz^2}f_\gamma(z) = -\sum_{\gamma'}\omega_{\gamma,\gamma'}(z)f_{\gamma'}(z) \quad (6.11)$$

where the matrix, $\omega_{\gamma\gamma'}(z)$ is written as

$$\omega_{\gamma,\gamma'}(z) = \sum_{\gamma''}\sum_{\gamma'''}(M^{-1})_{\gamma,\gamma''}(z)W_{\gamma'',\gamma'''}(z)M_{\gamma''',\gamma'}(z) \qquad (6.12)$$

$$W_{\gamma,\gamma'}(z) = k_{\gamma}^{2}(z)\delta_{\gamma,\gamma'} - \{C^{(0,1)}(z)\}^{2}_{\gamma,\gamma'} - C^{(1,1)}_{\gamma,\gamma'}(z) - \frac{2m^*}{\hbar^2}V^{\mathrm{IM}}_{\gamma,\gamma'}(z) \qquad (6.13)$$

$$C^{(1,1)}_{\gamma,\gamma'}(z) = \int dx \int dy \frac{\partial\varphi_{\gamma}(x,y|z)}{\partial z}\frac{\partial\varphi_{\gamma'}(x,y|z)}{\partial z} \qquad (6.14)$$

$$V^{\mathrm{IM}}_{\gamma,\gamma'}(z) = \int dx \int dy \varphi^{*}_{\gamma}(x,y|z)V_{\mathrm{IM}}(x,y,z)\varphi_{\gamma'}(x,y|z) \qquad (6.15)$$

and the expression, $\{C^{(0,1)}(z)\}^{2}_{\gamma,\gamma'}$, in eqn (6.13) represents the (γ,γ')-element of the multiplied matrix, $C^{(0,1)}(z) \cdot C^{(0,1)}(z)$.

A set of renormalised complex wavenumbers, $K_{\gamma}(z)$, which includes lateral-mode mixing, is obtained by solving the eigenvalue equation:

$$\sum_{\gamma''}W_{\gamma,\gamma''}(z)V_{\gamma'',\gamma'}(z) = \{K_{\gamma'}(z)\}^{2}V_{\gamma,\gamma'}(z) \qquad (6.16)$$

where $V_{\gamma\gamma'}(z)$ is a unitary matrix which diagonalises the matrix, $W_{\gamma\gamma'}(z)$. Then the z-component of the wavefunction can be expressed as a superposition of plane waves:

$$\begin{aligned}\chi_{\gamma}(z) = \sum_{\gamma'}\sum_{\gamma''}M_{\gamma,\gamma'}(z)V_{\gamma',\gamma''}(z)\{&A_{\gamma''}(z)\exp(\mathrm{i}K_{\gamma''}(z)z) \\ &+ B_{\gamma''}(z)\exp(-\mathrm{i}K_{\gamma''}(z)z)\}\end{aligned} \qquad (6.17)$$

where $A_{\gamma}(z)$ and $B_{\gamma}(z)$ are the coefficients of the forward and backward plane waves in the γth lateral mode with the complex wavenumber, $K_{\gamma}(z)$. Equation (6.16) is discretised on the finite-difference z-mesh points. Assuming these coefficients to be constant between two adjacent z-mesh points, the 3D wavefunction, $\Psi(x,y,z)$, can finally be written as

$$\Psi^{(i)}(x,y,z) \cong \sum_{\gamma}\sum_{\gamma'}\sum_{\gamma''}\varphi_{\gamma}(x,y|z)M^{(i)}_{\gamma,\gamma'}V^{(i)}_{\gamma',\gamma''}\{A^{(i)}_{\gamma''}\exp(\mathrm{i}K^{(i)}_{\gamma''}z) + B^{(i)}_{\gamma''}\exp(-\mathrm{i}K^{(i)}_{\gamma''}z)\} \qquad (6.18)$$

where the index, (i), denotes the small region between adjacent z-mesh points z_i and z_{i+1}.

From the continuity of electron probability flux of electrons through the system, the following conditions on the total wavefunctions hold at the z-mesh point z_{i+1}, for given x and y:

$$\Psi^{(i)}(x,y,z_{i+1}) = \Psi^{(i+1)}(x,y,z_{i+1}) \tag{6.19}$$

$$\frac{1}{m^*}\frac{\partial}{\partial z}\Psi^{(i)}(x,y,z)\bigg|_{z=z_{i+1}} = \frac{1}{m^*}\frac{\partial}{\partial z}\Psi^{(i+1)}(x,y,z)\bigg|_{z=z_{i+1}} \tag{6.20}$$

The coefficients at adjacent z-mesh points are then related as follows:

$$\begin{pmatrix} A_{\gamma}^{i+1} \\ B_{\gamma}^{i+1} \end{pmatrix} = \sum_{\gamma'} T^{(i)}(\gamma,\gamma') \begin{pmatrix} A_{\gamma'}^{i} \\ B_{\gamma'}^{i} \end{pmatrix} \tag{6.21}$$

The matrix, $T^{(i)}(\gamma,\gamma')$, can be expressed as

$$T^{(i)}(\gamma,\gamma') = \begin{pmatrix} \alpha_{+}^{(i)}(\gamma,\gamma')\exp\{i(K_{\gamma'}^{(i)} - K_{\gamma}^{(i+1)})z_{i+1}\} & \alpha_{-}^{(i)}(\gamma,\gamma')\exp\{-i(K_{\gamma'}^{(i)} + K_{\gamma}^{(i+1)})z_{i+1}\} \\ \alpha_{-}^{(i)}(\gamma,\gamma')\exp\{i(K_{\gamma'}^{(i)} + K_{\gamma}^{(i+1)})z_{i+1}\} & \alpha_{+}^{(i)}(\gamma,\gamma')\exp\{-i(K_{\gamma'}^{(i)} - K_{\gamma}^{(i+1)})z_{i+1}\} \end{pmatrix} \cdot X_{\gamma,\gamma'}^{(i)} \tag{6.22}$$

where the matrices, $\alpha_{\pm}^{(i)}(\gamma,\gamma')$ and $X_{\gamma,\gamma'}^{(i)}$, are given by the following expressions:

$$\alpha_{\pm}^{(i)}(\gamma,\gamma') = \frac{1}{2}\left(1 \pm \frac{m_{i+1}^*}{m_i^*}\frac{K_{\gamma'}^{(i)}}{K_{\gamma}^{(i+1)}}\right) \tag{6.23}$$

$$X_{\gamma,\gamma'}^{(i)} = \sum_{\gamma_1}\sum_{\gamma_2}\sum_{\gamma_3} V_{\gamma_1,\gamma}^{(i+1)} M_{\gamma_2,\gamma_1}^{(i+1)} M_{\gamma_2,\gamma_3}^{(i)} V_{\gamma_3,\gamma'}^{(i)} \tag{6.24}$$

Hence the coefficients at the emitter, $(A_{\gamma}^{L}, B_{\gamma}^{L})$, and collector, $(A_{\gamma}^{R}, B_{\gamma}^{R})$, edges of the device are related using a multi-mode transfer matrix, $T(\gamma,\gamma')$:

$$\begin{pmatrix} A_{\gamma}^{L} \\ B_{\gamma}^{L} \end{pmatrix} = \sum_{\gamma'} T(\gamma,\gamma') \begin{pmatrix} A_{\gamma'}^{R} \\ B_{\gamma'}^{R} \end{pmatrix} \tag{6.25}$$

$$T = T^{(N)}T^{(N-1)}T^{(N-2)}\ldots T^{(2)}T^{(1)} \tag{6.26}$$

It should be noted that the transfer matrix may contain both propagating and evanescent modes, depending on the total energy and lateral-mode eigenenergies. As long as we are interested in only the transport properties of the 0D RTD, the evanescent modes are not required, and a reduced transfer matrix [26] may be calculated from the above full transfer matrix which separates the propagating modes from the evanescent modes. In the present calculations, however, there is always a difference in the number of propagating modes at the emitter and collector edges under a non-zero external bias, and the resulting reduced transfer matrix is no longer regular. In the following calculations the full transfer matrix is adopted rather than the reduced transfer matrix [26]. A relevant multi-mode scattering matrix, $S(\gamma,\gamma')$, which is defined as

$$\begin{pmatrix} B^{L}_{\gamma} \\ A^{R}_{\gamma} \end{pmatrix} = \sum_{\gamma'} S(\gamma,\gamma') \begin{pmatrix} A^{L}_{\gamma'} \\ B^{R}_{\gamma'} \end{pmatrix} \tag{6.27}$$

is calculated from the transfer matrix. The multi-mode transmission probability, $t_{L,R}(E;\gamma,\gamma')$, and the total transmission rate, $T(E)$, that is, the conductance at zero temperature, are then obtained from the S-matrix as follows:

$$t_{R}(E;\gamma,\gamma') = |S_{12}(\gamma,\gamma')|^2 \tag{6.28}$$

$$t_{L}(E;\gamma,\gamma') = |S_{21}(\gamma,\gamma')|^2 \tag{6.29}$$

$$\begin{aligned} T(E) &= \sum_{\gamma}\sum_{\gamma'} t_{R}(E;\gamma,\gamma')\theta(E-\epsilon^{L}_{\gamma})\theta(E-\epsilon^{R}_{\gamma'}) \\ &= \sum_{\gamma}\sum_{\gamma'} t_{L}(E;\gamma,\gamma')\theta(E-\epsilon^{L}_{\gamma})\theta(E-\epsilon^{R}_{\gamma'}) \end{aligned} \tag{6.30}$$

where $\theta(E)$ is the step-function. A complete set of 3D wavefunctions, $\Psi(x,y,z)$, can be obtained by using the following scattering boundary conditions:

$$(A^{L}_{\gamma}, B^{R}_{\gamma}) = (\delta_{\gamma,\gamma_0}, 0) \qquad (\gamma = 1,2,3,\ldots) \tag{6.31}$$

for an incident electron wave with a lateral mode, γ_0, originating at the emitter edge of the system, and

$$(A^{L}_{\gamma}, B^{R}_{\gamma}) = (0, \delta_{\gamma,\gamma_0}) \qquad (\gamma = 1,2,3,\ldots) \tag{6.32}$$

when it originates at the collector edge. In eqns (6.31) and (6.32) δ_{γ,γ_0} is the delta function. If the system has a real bound state, which may be caused by an attractive scatterer such as an InGaAs quantum well or a deep donor trap level, a bound state problem has to be solved as well as the above scattering state problem to obtain the complete set of wavefunctions. Finally, the total tunnelling current, I_{tunnel}, is calculated assuming global coherent tunnelling of electron waves throughout the device as follows:

$$I_{tunnel} = \frac{e}{\pi\hbar}\int_{eV}^{\infty} T(E)\{f_L(E) - f_R(E)\}dE \tag{6.33}$$

where $f_L(E)$ and $f_R(E)$ are the Fermi distribution functions in the emitter and collector regions respectively.

6.2.2 Lateral-mode conserving and non-conserving resonant tunnelling

In this section the 3D scattering theory, described in Section 6.2.1, is applied to the 0D RTD structure shown in Fig. 6.6, and multi-mode resonant tunnelling is analysed numerically. The hour-glass-shaped confinement potential due to surface Fermi-level pinning is calculated in advance using a classical device simulation with Spicer's surface defect model [27],[28]. The calculated lateral confinement potential, $V_{LC}(x,y,z)$, is then introduced into eqn (6.2). We focus on the effect of lateral-mode mixing, caused by elastic scattering due to the hour-glass-shaped confinement potential, on the multi-mode transmission properties and I–V characteristics of the 0D RTD. The S-matrix is calculated to analyse lateral-mode non-conserving tunnelling which can be observed in the off-diagonal components of the transmission probability. Furthermore, a total tunnelling current through the device is calculated and compared with results for a device with uniform lateral confinement in order to investigate the mechanism of the fine structure in the I–V characteristics.

In the present calculations we model a laterally confined AlGaAs/InGaAs/AlGaAs double-barrier resonant tunnelling structure. The assumed layer structure consists of an undoped $In_{0.08}Ga_{0.92}As$ quantum well 5 nm thick, two undoped $Al_{0.25}Ga_{0.75}As$ barriers 4 nm thick, two undoped GaAs spacer layers 6 nm thick, and n^+-type GaAs emitter and collector layers with donor concentrations of 1.0×10^{18} cm^{-3}. The conduction-band discontinuities in the GaAs/$Al_{0.25}Ga_{0.75}As$ and GaAs/$In_{0.08}Ga_{0.92}As$ heterostructures are assumed to be 187.0 and -37.2 meV respectively. The electron effective mass in the GaAs, $Al_{0.25}Ga_{0.75}As$ and $In_{0.08}Ga_{0.92}As$ layers is assumed to be 0.067, 0.088 and 0.064 m_0 respectively. The lateral dimensions of the device are set to be 80 nm in both the x- and y-dimensions.

The first step of the numerical simulation is to obtain a realistic lateral confinement potential distribution created by the carrier trap levels on the lateral surface. The exact lateral confinement potential should be determined through a fully self-consistent calculation of the 3D Schrödinger equation. However, this would require an enormous amount of computational time and is beyond our present requirements. The self-consistent calculations are left for a future analysis of Coulomb blockade tunnelling, where the self-consistent field produced by a single electron is dominant. In this section the confinement potential is calculated using a classical device simulation [27] in which surface carrier traps are taken into consideration using Spicer's unified defect model [28]: a deep donor level at 0.925 eV measured from the conduction-band edge and a deep acceptor level at 0.75 eV from the valence-band edge are assumed for the GaAs lateral surface. As long as the size of the lateral confinement is much larger than the width of the quantum well, the calculated potential distribution should be a fairly good approximation to the exact potential distribution determined by self-consistent calculation. Figure 6.6 shown in Section 6.1 is the calculated potential distribution where the sheet concentration of the surface deep level is assumed to be 5.0×10^{12} cm^{-2}; a plausible value which is large enough to pin the Fermi-level on the surface. It can be seen that the hour-glass-shaped confinement potential results from the different surface depletion widths in the intrinsic and contact regions.

The second step of the simulation is to calculate the lateral eigenenergies and 2D eigenstates at each z-point using the hour-glass confinement potential. Equation (6.4) is discretised by using a 3D finite-difference lattice, shown in Fig. 6.8, which has a uniform mesh spacing in the x- and y-dimensions and a non-uniform spacing in the z-dimension.

Eigenenergies of the resultant finite-difference equation are obtained up to a given value of total energy by using the bisection method following Householder's tridiagonalisation. The corresponding eigenvectors are then calculated by the inverse iteration method [29]. To speed up the search for the eigenvectors, the set of eigenstates obtained at the previous z-mesh point is used as an initial guess for the eigenstates at the next z-mesh point. For numerical calculations a cut-off value is introduced for the maximum eigenenergy, although all the lateral modes would be necessary to form a complete set. The number of wavefunctions required for realistic calculations depends on the system under

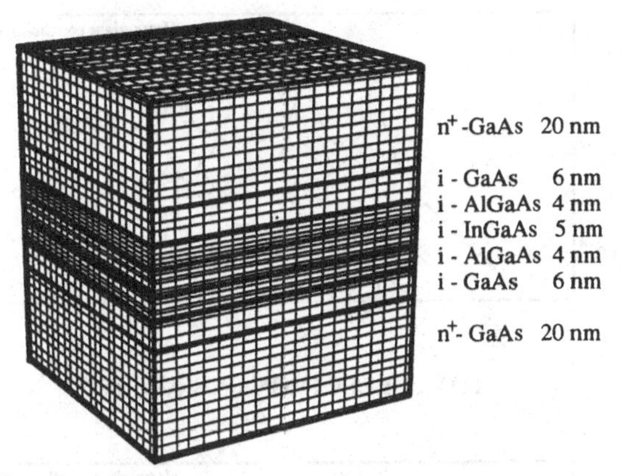

Figure 6.8 3D finite-difference lattice used for numerical calculations. The mesh spacing has been chosen to be small for the double-barrier structure.

consideration. In general, at least all of the lateral eigenstates with eigenenergies below the local Fermi energy in the emitter should be taken into account. In the present calculations, for instance, there are four independent lateral eigenstates below the Fermi level and 13 lateral modes are calculated for all z-mesh points.

By making use of the lateral eigenstates, the mixing coefficients, $C^{(0,1)}_{\gamma,\gamma'}$ and $C^{(0,2)}_{\gamma,\gamma'}$, and the unitary transformation matrix, $M_{\gamma\gamma'}(z)$, can be evaluated from eqns (6.7), (6.8) and (6.10). The eigenvalue equation, eqn (6.16), is then solved for the renormalised wavenumbers, $K_\gamma(z)$, and the unitary matrix, $V_{\gamma\gamma'}(z)$. As the matrix, $W_{\gamma\gamma'}(z)$, is real and symmetric, all the eigenvalues and eigenvectors can be obtained using the QL method [29]. Finally, by using the lateral eigenstates at the emitter and collector edges, the multi-mode transfer matrix and the resultant scattering matrix are calculated from eqns (6.21)–(6.29).

The multi-mode transmission probability calculated for the 0D RTD structure is shown in Fig. 6.9 [30]. Figure 6.9(a) shows the total energy dependence of the transmission probability for diagonal tunnelling from the γth-incident-mode to the γth-transmission-mode, $|S_{12}(\gamma,\gamma)|^2$, and Fig. 6.9(b) shows that for the off-diagonal tunnelling from the first-incident-mode to the γth-transmission-mode, $|S_{12}(1,\gamma)|^2$. The S-matrix elements are drawn for values of γ up to 11. The transmission probability calculated assuming a completely uniform confinement is shown in Fig. 6.10 for comparison. Before proceeding to the detailed discussion about these results, it is worth commenting on this structure.

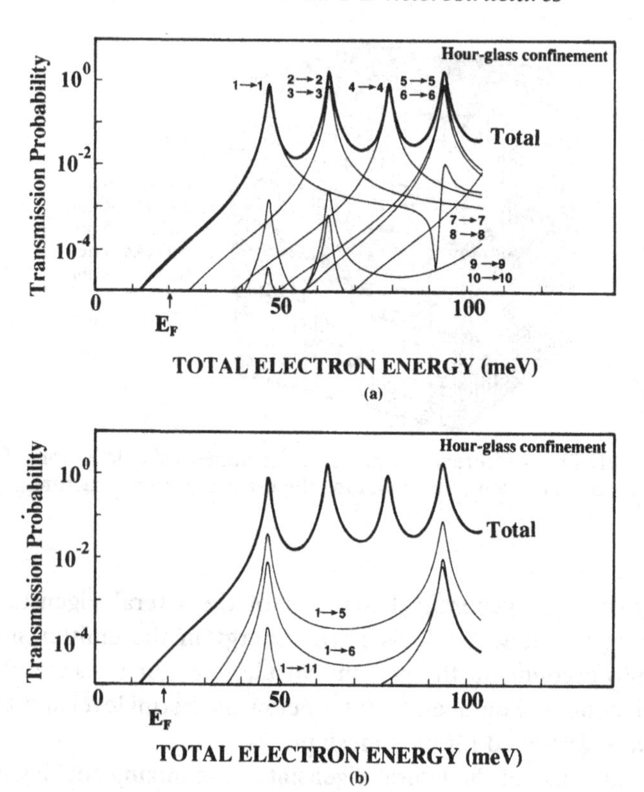

Figure 6.9 Multi-mode transmission probability calculated for a 0D RTD with hour-glass lateral confinement: (a) total energy dependence of the transmission probability for diagonal tunnelling from the γth-incident mode to the γth-transmission mode, $|S_{12}(\gamma,\gamma)|^2$, and (b) that for off-diagonal tunnelling from the first-incident mode to the γth-transmission mode, $|S_{12}(1,\gamma)|^2$. The *S*-matrix elements are drawn for γ up to 11. The bold solid line represents the total transmission probability $T(E)$ defined by eqn (6.30). (E_F is the Fermi energy.)

The uniform lateral confinement used for the calculation in Fig. 6.10 could be realised by uniformly doping the resonant tunnelling structure as well as the contact regions. As we saw in Section 3.3, tunnelling electrons suffer from frequent impurity scattering in a doped quantum well, resulting in collisional broadening which degrades the current *P*/*V* ratio. Thus the results illustrated in Fig. 6.10 are based on an unrealistic assumption that electron waves travel in a coherent manner even in a doped tunnelling structure, and are given simply for comparison with the results for the hour-glass confinement.

In Fig. 6.10 the *S*-matrix has no off-diagonal elements since there is no lateral wavefunction mismatch anywhere in the system. This is called

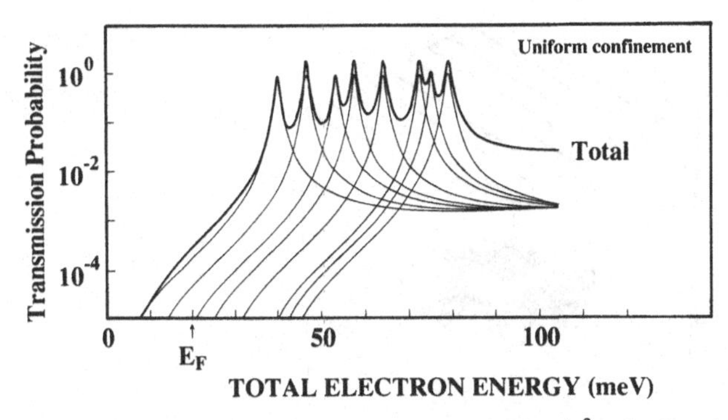

Figure 6.10 Multi-mode transmission probability $|S_{12}(\gamma,\gamma)|^2$ calculated for a 0D RTD with uniform lateral confinement. The off-diagonal elements are now zero. (E_F is the Fermi energy.)

independent-mode tunnelling, as the lateral modes are not mixed at all. The total transmission rate is then just a superposition of the transmission probabilities through these independent modes, shown as a bold solid line in Fig. 6.10. The electronic states corresponding to the first three transmission probability peaks in Fig. 6.10 are shown in Fig. 6.11(a)–(d). This diagram shows the visualised 3D probability density of electrons, $|\Psi_E(x,y,z)|^2$, at resonance: (a) for the first-mode incident wave from the emitter at the first peak energy, (b) for the second-mode incident wave from the emitter at the second peak energy, (c) for the third-mode incident wave from the emitter at the second peak energy, and (d) for the fourth-mode incident wave from the emitter at the third peak energy. It should be noted that the wavefunctions for these states are virtually localised in the quantum dot and clearly reflect the eigenstates of the quantum dot itself. This fact means that the lateral mode index, γ, is a good quantum number throughout the device.

By contrast, the following two major differences can be seen in the tunnelling properties of the hour-glass confinement shown in Fig. 6.9. First, the energy intervals between the transmission probability peaks become larger than those in Fig. 6.10, leading to a large peak-to-valley ratio of transmission probability. Second, elastic scattering due to the hour-glass confinement potential mixes the lateral modes and opens new off-diagonal tunnelling channels. In Fig. 6.9(b) two peaks can be found in the off-diagonal elements of the S-matrix which represent lateral-mode non-conserving resonant tunnelling.

It should be noted that off-diagonal tunnelling with the first-incident mode is observed only for the fifth-, sixth- and eleventh-transmission

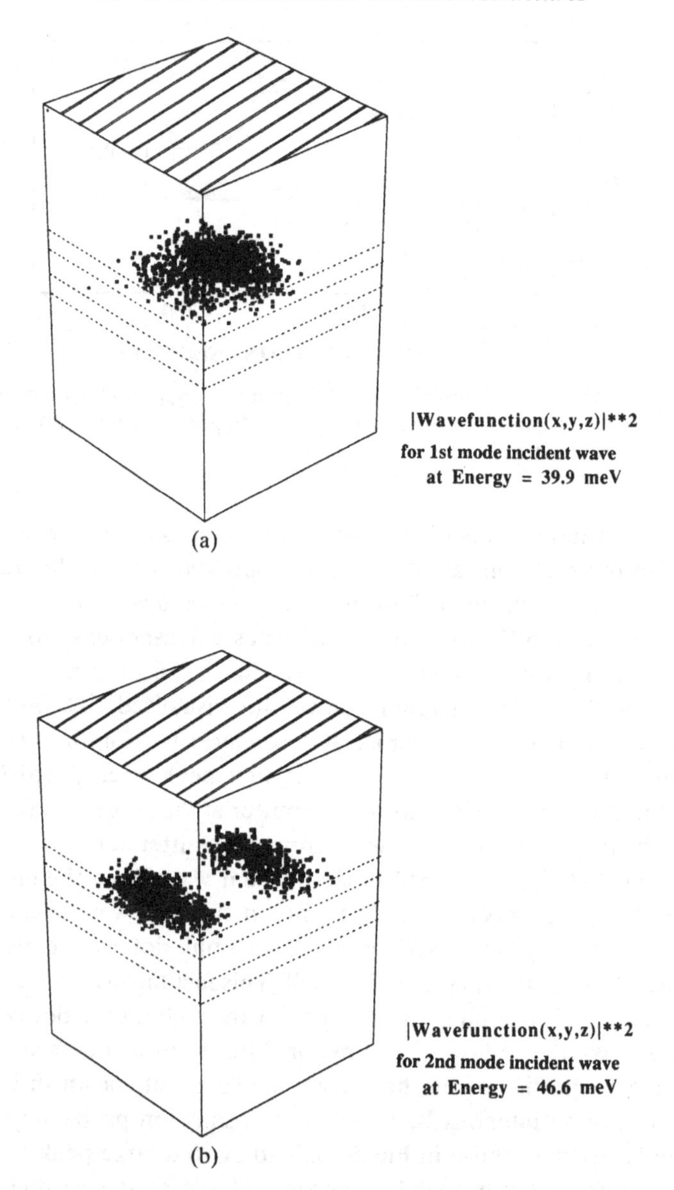

Figure 6.11 Visualised 3D probability density of electrons, $|\Psi_E(x,y,z)|^2$, at the lowest three transmission peaks in Fig. 6.10: (a) for the first-incident mode at the first peak energy, (b) for the second-incident mode at the second peak energy, (c) for the third-incident mode at the second peak energy, and (d) for the fourth-incident mode at the third peak energy. It should be noted that the second peak in Fig. 6.10 is caused by both tunnelling processes $2 \rightarrow 2$ and $3 \rightarrow 3$. It can be seen that the electron distribution clearly reflects the lateral eigenstates in the quantum dot.

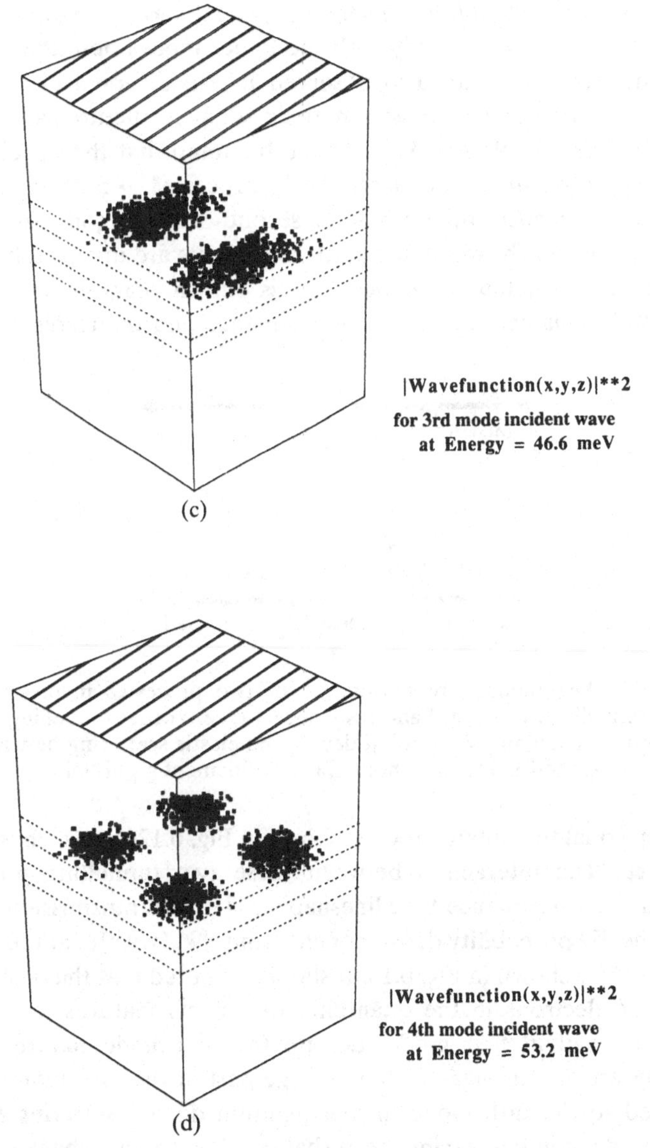

Figure 6.11 *cont.*

modes. This is purely because of a selection rule for parity of lateral wavefunctions. Since the elastic scattering due to the hour-glass confinement potential does not break symmetry under mirror reflection in the x- and y-dimensions, a lateral mode only couples with other modes having the same parity.

The lowest wavefunction has even parity in both x- and y-dimensions, and can therefore couple only with the upper modes described above. Additional structures are also observed in the diagonal elements in Fig. 6.9(a). For example, an asymmetric resonant structure can be seen at a total energy of 91.7 meV. It should be noted that the off-diagonal tunnelling probability is quite large for the process, $1 \rightarrow 5$, at this energy. In these circumstances the ratio of the second-order diagonal tunnelling is enhanced in which two elastic scattering events are involved between the first and fifth lateral modes: this is shown diagrammatically in Fig. 6.12(b). This process is now at resonance and interferes with the

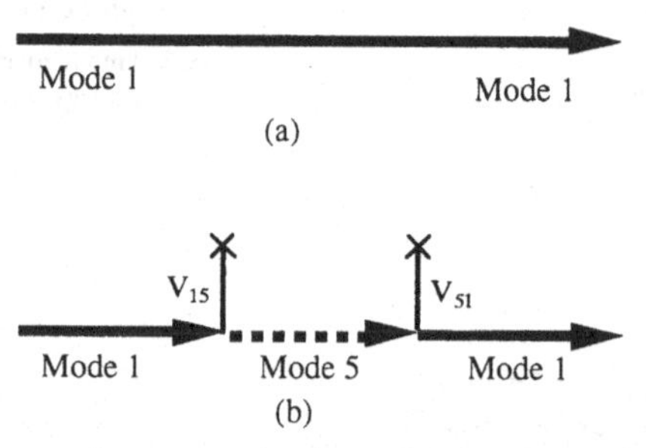

Figure 6.12 Diagrammatic representation of two processes involved in the diagonal tunnelling, $1 \rightarrow 1$, at Fano-resonance: (a) zero order tunnelling and (b) second order tunnelling. V_{51} and V_{15} denote the elastic scattering between first and fifth lateral modes due to an hour-glass confinement potential.

major diagonal tunnelling process, shown in Fig. 6.12(a), which is at off-resonance. The interaction between these two tunnelling processes results in Fano-resonance-type lineshape [31] in the transmission probability. The 3D probability density of electrons, $|\Psi_E(x,y,z)|^2$, at the energy of 91.7 meV is shown in Fig. 6.13. It should be noted that the probability density of electrons in the quantum dot reflects features of the fifth mode rather than first mode, despite the first-mode nature of the incoming wave. This signifies that a large part of the incoming wave is converted to the fifth mode in the quantum dot by suffering from a lateral-mode non-conserving perturbation due to the change in the lateral confinement. In these circumstances the lateral-mode index, γ, is no longer an appropriate quantum number for the whole system.

The applied voltage dependence of the total tunnelling current is calculated by assuming global coherent tunnelling throughout the device (eqn (6.33)). Figure 6.14(a) shows the I–V characteristics of a device with hour-glass lateral confinement, calculated at a temperature of 77 K.

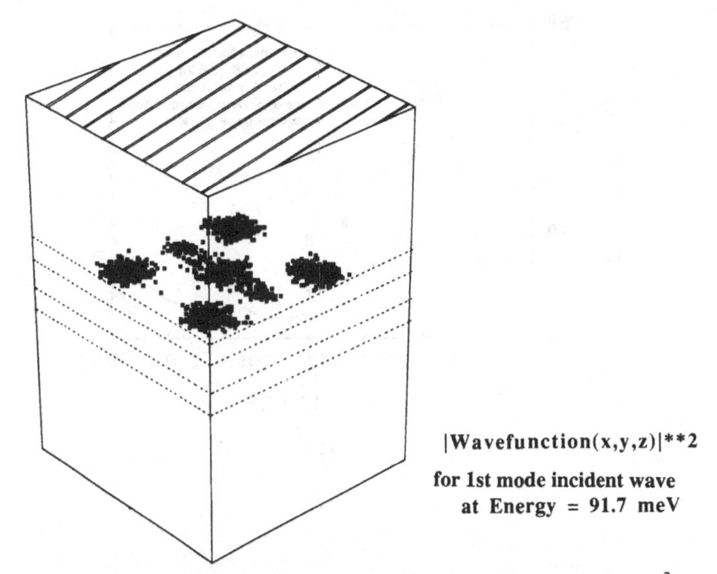

Figure 6.13 Visualised 3D probability density of electrons, $|\Psi_E(x,y,z)|^2$, for the first-incident mode at an energy of 91.7 meV which corresponds to the dip of the Fano-resonance in Fig. 6.9. It should be noted that the probability density of electrons in the quantum dot reflects the features of the fifth mode rather than the first mode, despite the first-mode nature of the incoming wave.

For an applied bias a piecewise-linear model has been adopted for the external potential $V_{EX}(z)$ in eqn (6.2), which was introduced in Chapter 2 for large-area RTDs: a uniform external electric field is assumed in the intrinsic regions of the device. Several satellite current peaks and shoulders are observed superposed on the conventional NDC characteristics of 2D RTDs. Figure 6.14(b) shows the *I–V* characteristics calculated at lower temperatures. The fine structure in the *I–V* characteristics can be seen more clearly at lower temperatures.

In Fig. 6.15, the *I–V* characteristics are compared with those of the device with uniform confinement (broken line). It should be noted that, at a temperature of 77 K, only one major current peak is found in the case of the uniform confinement. No fine structure is seen since the peak-to-valley ratio of the transmission probability shown in Fig. 6.10 is not large enough to separate the contribution from each mode at this temperature. At lower temperatures peaks and shoulders can clearly be seen which correspond exactly to transmission peaks, as indicated in Fig. 6.15(b).

In the case of hour-glass confinement, on the other hand, the

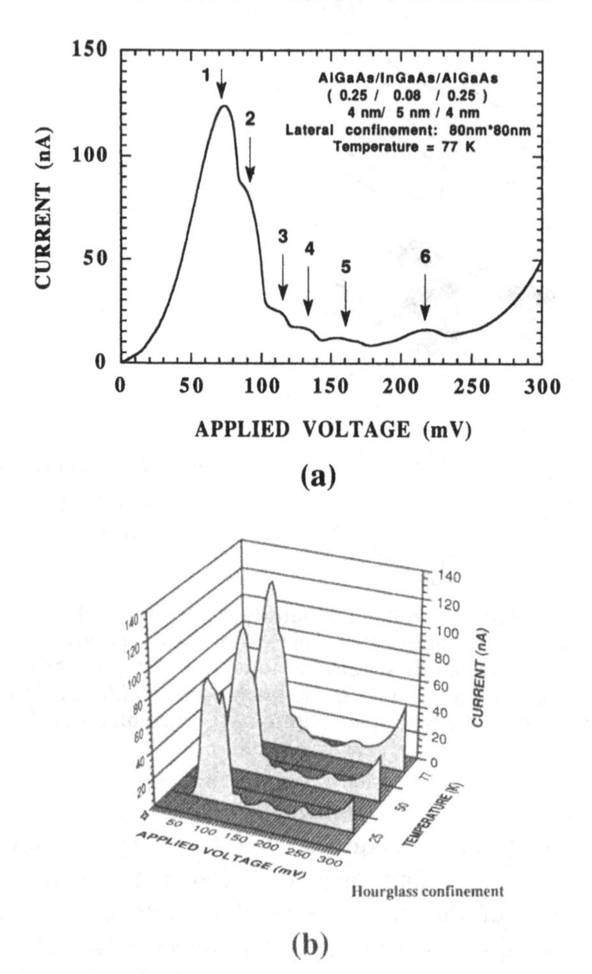

Figure 6.14 (a) Applied voltage dependence of total tunnelling current calculated using eqn (6.33) at a temperature of 77 K: observed satellite current peaks and shoulders are indicated by arrows. (b) *I–V* characteristics calculated at lower temperatures.

mechanism behind the fine structure seen in Fig. 6.14 is found to be more complicated, as explained below. The total energy dependence of the transmission probabilities calculated at the first four peak (shoulder) voltages are shown in Fig. 6.16(a)–(h): the diagonal elements, $|S_{12}(\gamma,\gamma)|^2$, calculated at the first, second, third and fourth peak voltages in Fig. 6.16(a), (c), (e) and (g) and the off-diagonal elements, $|S_{12}(1,\gamma)|^2$, in Fig. 6.16(b), (d), (f) and (h). The total transmission rate is shown by bold solid lines, as in Figs. 6.9 and 6.12. A current peak (shoulder)

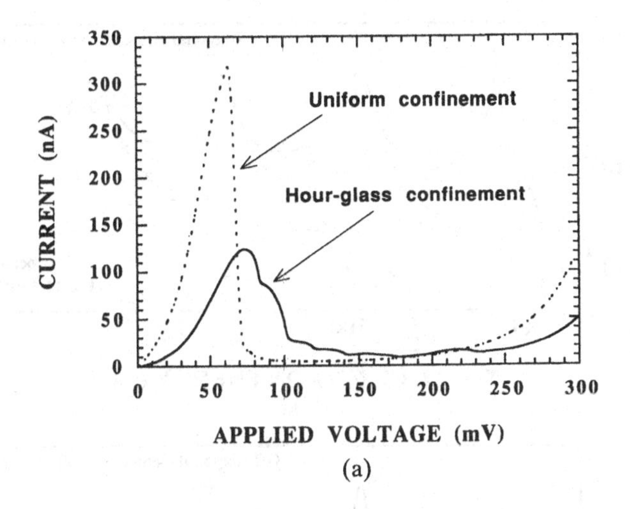

(a)

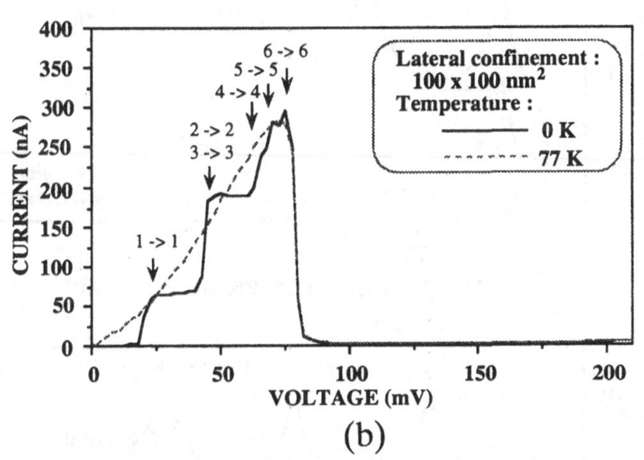

(b)

Figure 6.15 (a) Comparison of I–V characteristics calculated at 77 K for hourglass confinement (solid line) with those for uniform confinement (broken line). (b) The I–V characteristics calculated at $T = 0$ K for a uniform lateral confinement of 100 nm × 100 nm.

appears when each new transmission peak (indicated by an arrow) plunges into the Fermi sea. It can be seen that all four peaks are caused mainly by the lateral mode conserving tunnelling (Fig. 6.16(a), (c), (e) and (g)) since the lateral-mode non-conserving tunnelling shown in Fig. 6.16(b), (d), (f) and (h) contributes much less to the total transmission probability. Thus these four peaks mainly result from lateral-mode conserving resonant tunnelling: 1 → 1 tunnelling for the first main peak,

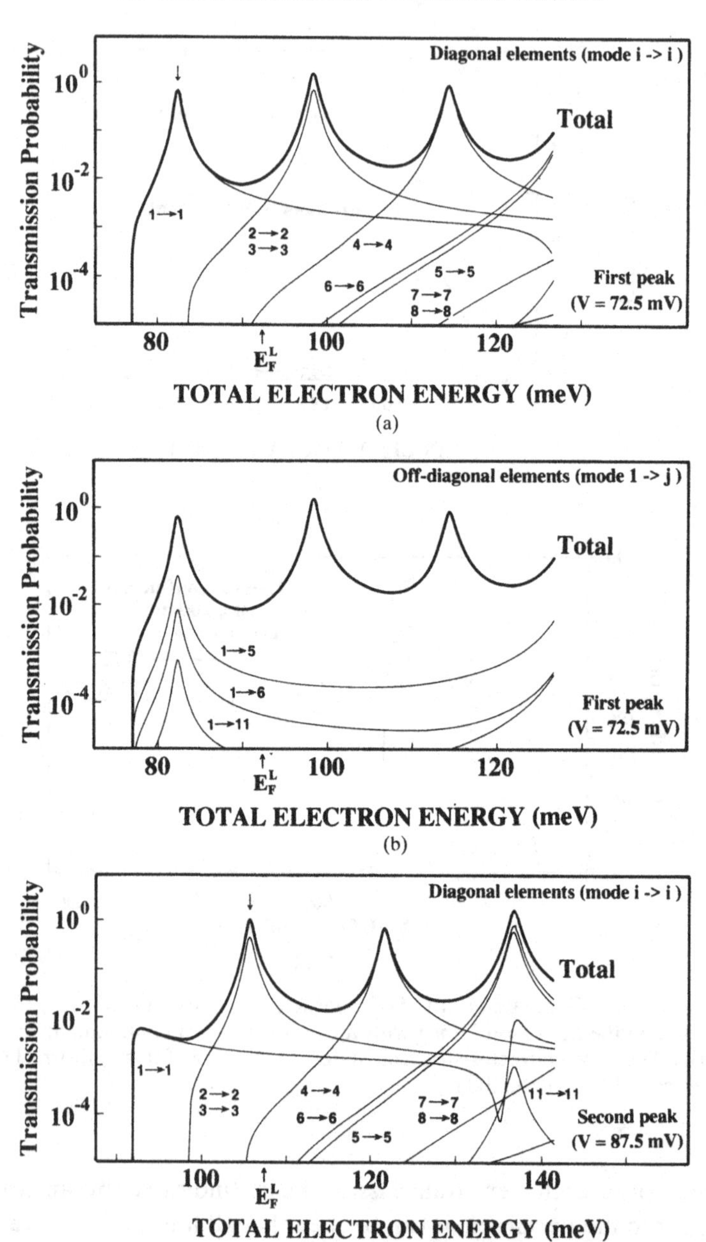

Figure 6.16 Total energy dependence of the transmission probabilities calculated at the first four peak (shoulder) voltages indicated by arrows 1, 2, 3 and 4 in Fig. 6.14: (a), (c), (e) and (g) show the diagonal elements, $|S_{12}(\gamma,\gamma)|^2$, and (b), (d) (f) and (h) show the off-diagonal elements, $|S_{12}(1,\gamma)|^2$. The total transmission rate is shown using bold solid lines, as in Figs. 6.9 and 6.10. E_F^L is the local Fermi energy in the emitter.

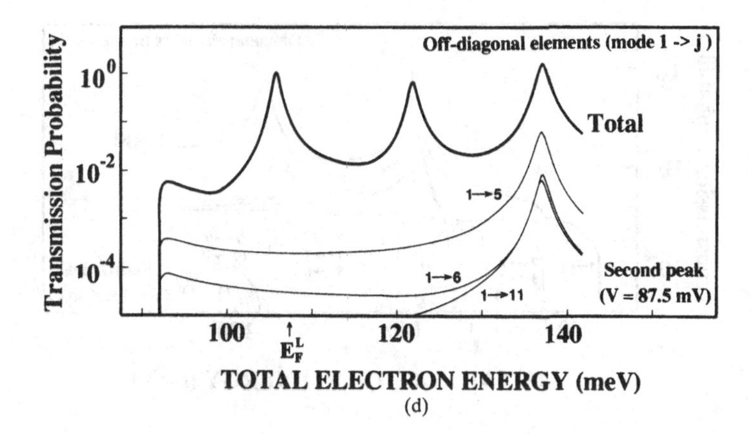

(d)

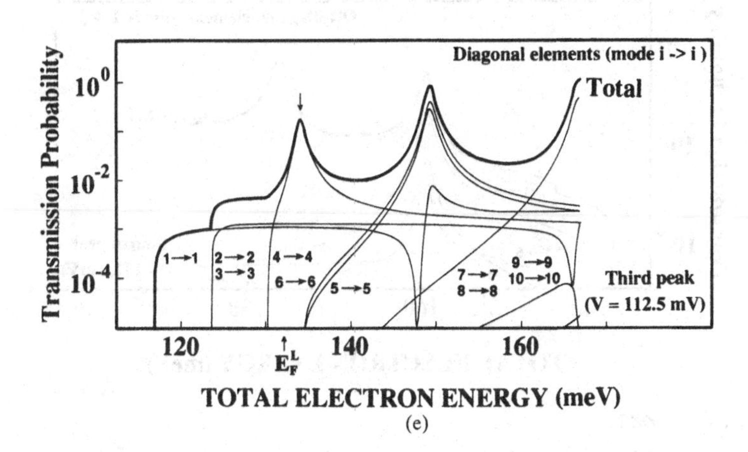

(e)

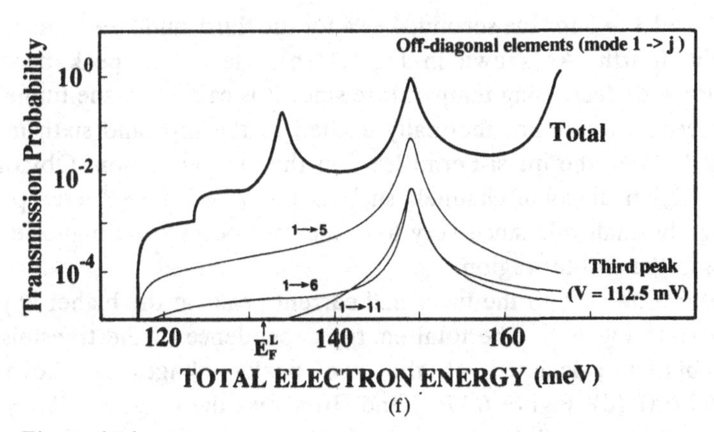

(f)

Figure 6.16 *cont.*

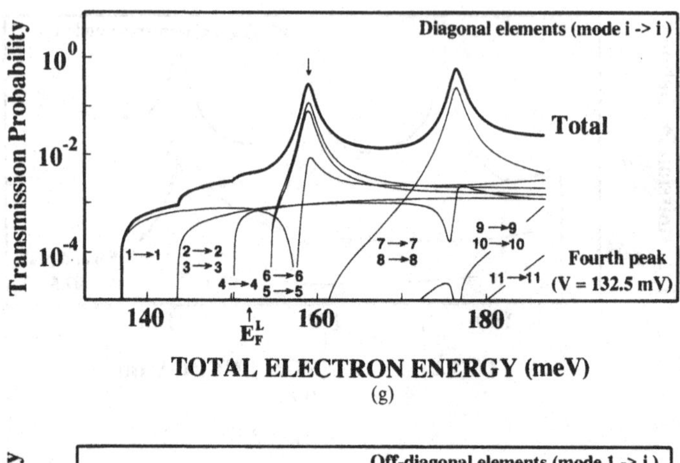

(g)

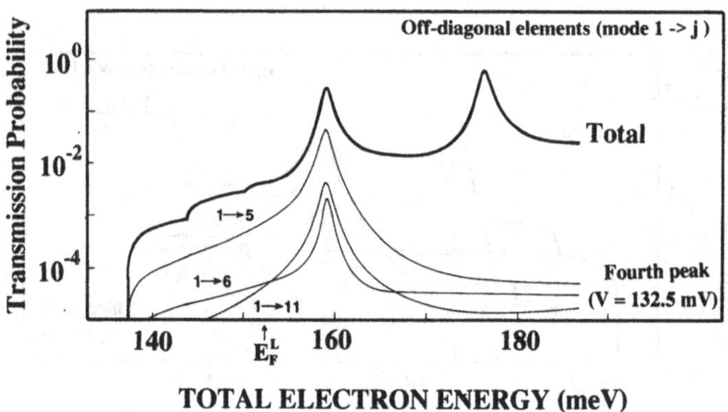

(h)

Figure 6.16 *cont.*

2 → 2 and 3 → 3 for the second, 4 → 4 for the third, and 5 → 5 and 6 → 6 for the fourth. As shown in Fig. 6.14(b), the fourth peak becomes smaller with decreasing temperature since it is caused by the tunnelling of electrons which are thermally excited to the fifth and sixth modes located above the quasi-Fermi level in the emitter region. Obviously, other higher-diagonal channels such as the 7 → 7, 8 → 8, etc., play a negligibly small role since very few electrons occupy the higher eigenstates in the emitter region.

Let us now turn to the two small current peaks at the higher applied voltages in Fig. 6.14. The total energy dependence of the transmission probabilities calculated at the two peak voltages is shown in Fig. 6.17(a)–(d). Figure 6.17(a) and (b) shows the diagonal, $|S_{12}(\gamma,\gamma)|^2$, and off-diagonal, $|S_{12}(1,\gamma)|^2$, at the fifth peak voltage, and Fig. 6.17(c)

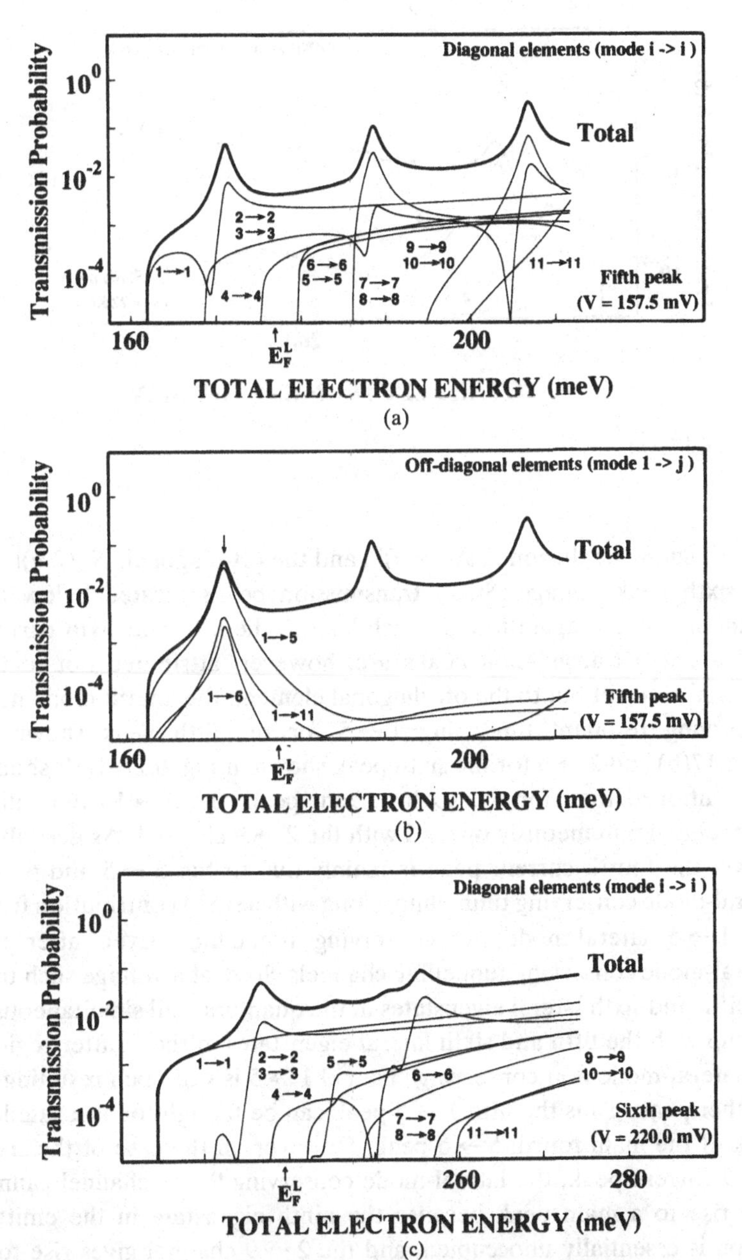

Figure 6.17 Total energy dependence of the transmission probabilities calculated at the last two peak voltages indicated by arrows 5 and 6 in Fig. 6.14: (a) and (c) show the diagonal elements, $|S_{12}(\gamma,\gamma)|^2$, (b) shows the off-diagonal, $|S_{12}(1,\gamma)|^2$, and, overleaf, (d) shows the off-diagonal, $|S_{12}(2,\gamma)|^2$. E_F^L is the local Fermi energy in the emitter.

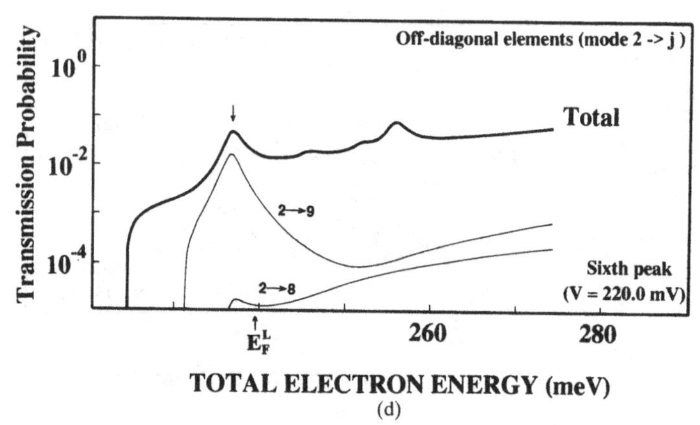

Figure 6.17 *cont.*

and (d) show the diagonal, $|S_{12}(\gamma,\gamma)|^2$, and the off-diagonal, $|S_{12}(2,\gamma)|^2$, at the sixth peak voltage. Small transmission peaks located below the Fermi energy are again found which lead to the fifth and sixth current peaks. These transmission peaks are, however, attributed not to the diagonal element but to the off-diagonal element: the lateral-mode non-conserving resonant tunnelling $1 \rightarrow 5$ for the fifth peak shown in Fig. 6.17(b) and $2 \rightarrow 9$ for the sixth peak shown in Fig. 6.17(d). It should be mentioned that, for the sixth current peak, the $3 \rightarrow 10$ tunnelling channel is simultaneously opened with the $2 \rightarrow 9$ channel. As described above, the fourth current peak is mainly due to the $5 \rightarrow 5$ and $6 \rightarrow 6$ lateral-mode conserving tunnelling, along with a small contribution from the $1 \rightarrow 5$ lateral-mode non-conserving tunnelling. Even after the lateral-mode conserving tunnelling channels close, at a voltage such that the fifth and sixth lateral eigenstates in the quantum well simultaneously line up with the fifth and sixth lateral eigenstates in the emitter region, the lateral-mode non-conserving channel $1 \rightarrow 5$ is still open resulting in another peak. Thus the fifth $1 \rightarrow 5$ peak can be thought of as a satellite peak of the main fourth $5 \rightarrow 5$ peak. However, in the case of the sixth $2 \rightarrow 9$ current peak, the lateral-mode conserving $9 \rightarrow 9$ channel cannot give rise to a main peak because the ninth eigenstate in the emitter region is essentially unoccupied, and the $2 \rightarrow 9$ channel gives rise to a new peak.

As shown here, the number of satellite current peaks (shoulders) observed in the *I–V* characteristics directly measures the number of resonant tunnelling channels in which the lateral mode is not conserved.

It may also be possible to estimate the magnitude of lateral-mode mixing by analysing the satellite peak current.

6.3 Gated resonant tunnelling structures – squeezable quantum dots*

6.3.1 Variable-area resonant tunnelling devices

As seen in Section 6.2.2, the hourglass-shaped confinement potential in an etched structure results in lateral-mode mixing and makes the characteristics of the device difficult to understand. The ideal lateral confinement that we want is one which is virtually flat around the tunnelling barriers (see Fig. 6.18) and so does not cause complicated

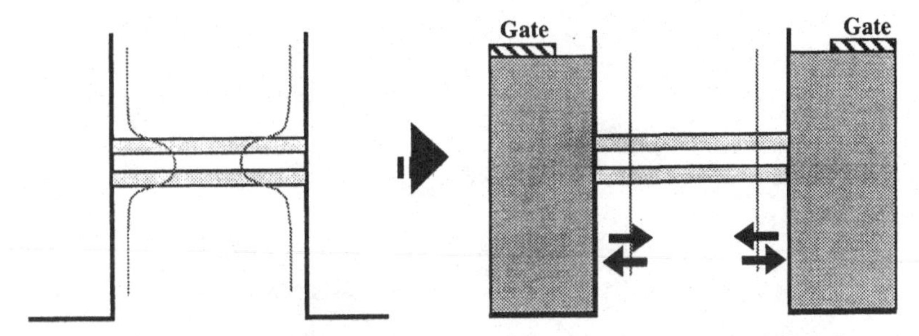

Figure 6.18 Ideal pseudo-uniform confinement potential controlled by gate electrodes.

mode mixing. As shown in Section 6.2.2, independent-mode tunnelling would be achieved if pseudo-uniform confinement were realised. In addition, it is useful if the size of the confinement is controllable for systematic investigations of 3D confinement.

For this purpose a new variable area resonant tunnelling diode (VARTD) [32]–[35] has been fabricated using Materials 2 and 3. Details of the tunnelling barrier structures in Materials 2 and 3 can be found in Fig. 2.17. Cross-sectional and perspective views of the VARTD are shown in Fig. 6.19(a) and (b), respectively. In the VARTD, lateral confinement arises from a reverse biased p–n-junction which is formed by using focused beryllium ion beam implantation (at 60 keV with an areal dose of 1×10^{14} cm^{-2}) after an anisotropic etch of depth 800 nm above the barriers.

* This section and Section 6.4 are based on the PhD dissertation by Goodings [33].

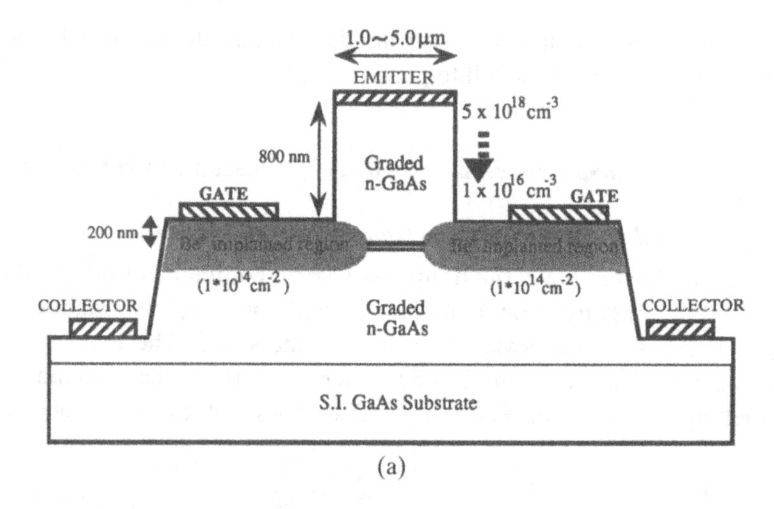

(a)

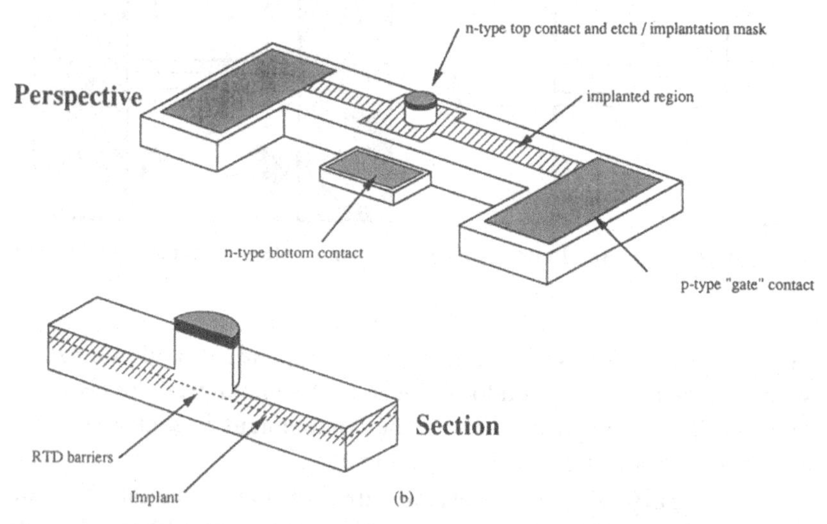

(b)

Figure 6.19 (a) Schematic cross-sectional and (b) perspective views of a VARTD.

As we have seen, both Materials 2 and 3 employ a graded doping profile in which the concentration of Si is reduced from 5×10^{18} cm^{-3} adjacent to the contacts to 1×10^{16} cm^{-3} near the barriers in order to achieve a large variation in depletion width. The thickness of the layer remaining above the barriers is adjusted to be about 200 nm so that the peak concentration of the implanted Be$^+$ can be aligned to the quantum well, resulting in a symmetric confinement potential. The diameter of

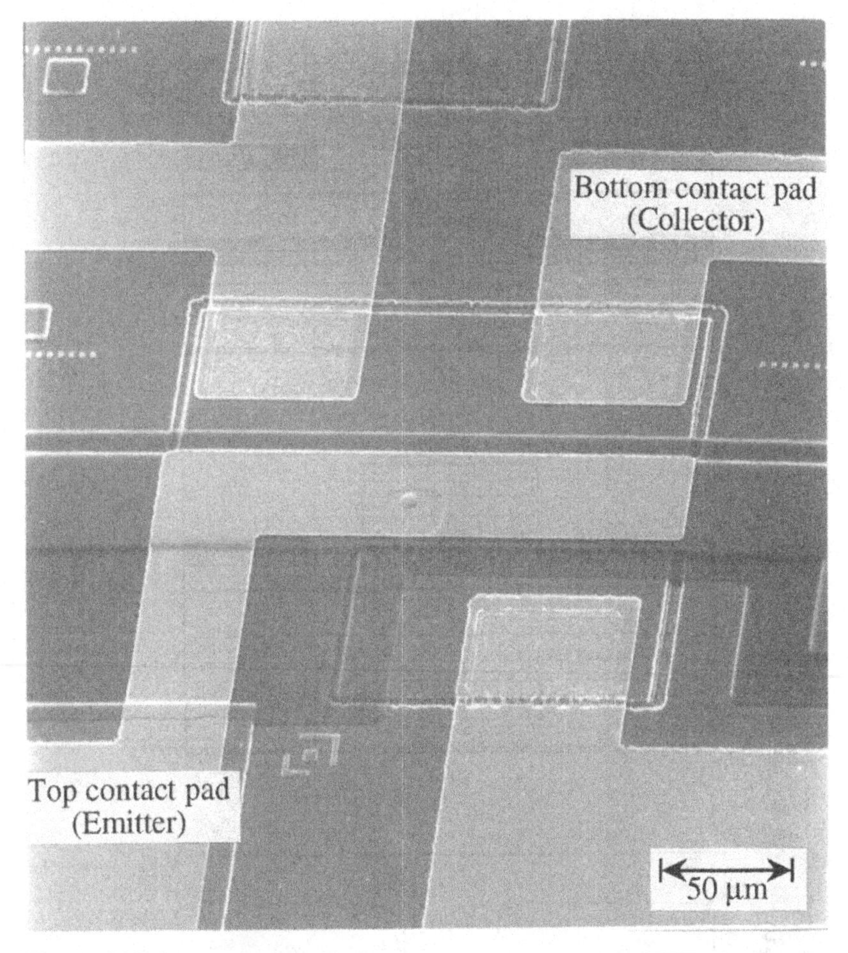

Figure 6.20 A perspective micrograph of a completed VARTD device. The small dot seen at the centre of the photograph is the device area. After Goodings [33], with permission.

the top contact pad is varied from 5 μm down to 1 μm. A perspective micrograph of a completed device is shown in Fig. 6.20 [33]: the VARTD has been planarised using a thin dielectric layer and an interconnect to the top contact has been deposited. The small dot seen in the centre of the diagram is the single VARTD.

Potential and current distributions have been calculated for a VARTD with a large-area contact pad by using a classical hydrodynamic device simulation [27] in order to illustrate the effects of graded

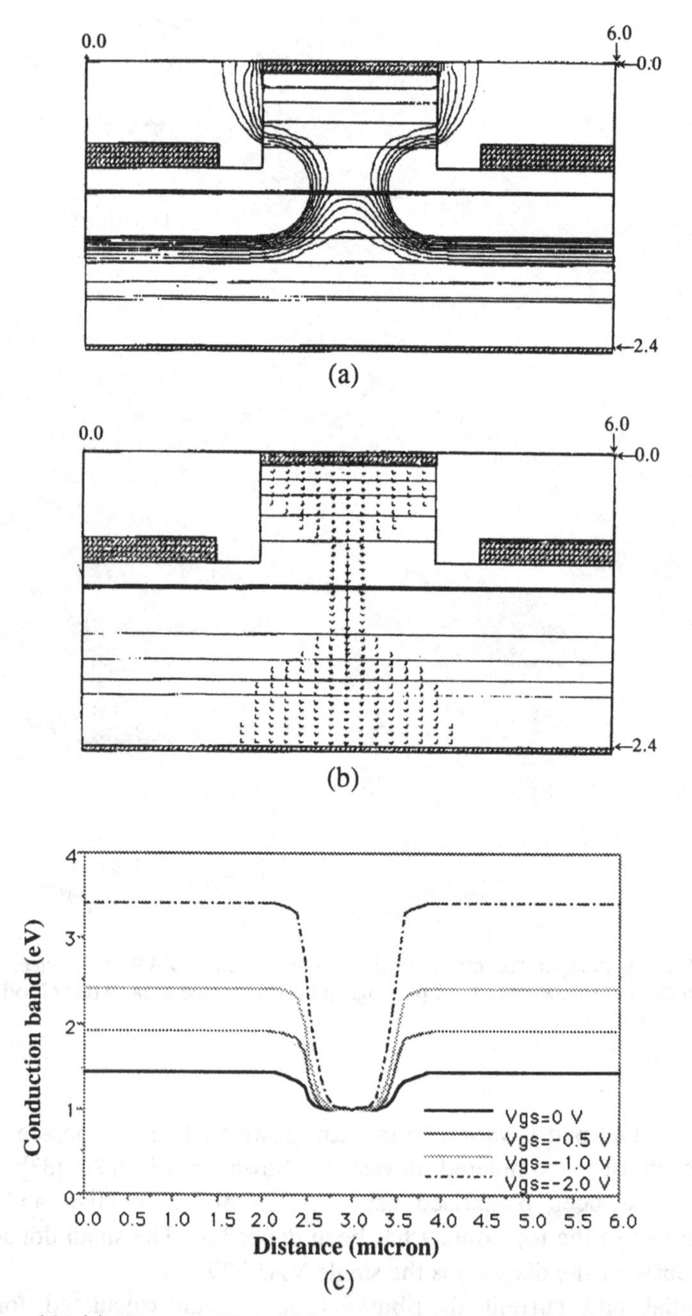

Figure 6.21 (a) 2D potential and (b) current distributions at a gate bias of −2.0 V and emitter–collector voltage of 1.0 V, and (c) lateral confinement potential profile for various values of gate bias calculated for a 2 μm device.

doping on the shape of the lateral confinement. Figure 6.21(a)–(c) shows (a) the potential distribution, (b) the current distribution and (c) the lateral confinement potential wells calculated for a 2 μm VARTD at a gate voltage of −2.0 V and an emitter–collector voltage of 1.0 V. In Fig. 6.21(a) and (b), lateral confinement turns out to be virtually flat around the barriers, as required. This results from the alignment of the peak concentration of implanted ions to the double-barrier structure and enables us to reduce the complicating effects of the asymmetric confinement potential which have been reported for a variable-area RTD with a surface Schottky gate [36]–[38].

We see from Fig. 6.21(c) that, as expected for a device of this size, the width of the lateral confinement potential well is far from the region in which lateral quantisation would appear and a smaller-area VARTD is necessary to observe the lateral confinement effects. In the following, however, we first analyse VARTDs with 2 μm and 3 μm top contact pads which are thought to be in an intermediate regime between conventional 2D RTDs, with complete lateral translational invariance, and three-dimensionally confined 0D RTDs. We show the squeezing effects of the applied gate bias on the *I–V* characteristics. The results on 3D confinement effects obtained from smaller VARTDs are presented in Section 6.3.2.

First let us illustrate the characteristics for the large-area VARTDs. Devices with various-sizes of top contact pad (from 2 μm to 5 μm in diameter) have been fabricated on Material 2. Typical *I–V* characteristics for a 3 μm diameter VARTD at 4.2 K are shown in forward- and reverse-bias directions in Fig. 6.22. With increasing negative gate bias the peak current can be reduced by a factor of approximately 2. The range of the negative gate bias is limited by reverse breakdown of the p–n-junction, which occurs at about −7 V. Hence, full pinch-off of the device, where the effects of lateral quantisation are expected to be observable, cannot be seen in this 3 μm device.

It should be noted that a small shift is seen in the resonant voltage for the 3 μm device (Fig. 6.22), and it is more pronounced for a 2 μm device (see Fig. 6.23) [33]–[35]. For devices larger than 3 μm this shift is not noticeable and it becomes greater as the device size is further reduced, as shown in Section 6.3.2. The shift of the resonant voltage towards a larger bias indicates that the gate depletion region induced by the reverse-biased p–n-junction begins to affect the potential energy of the channel, and the system is under transition from 2D to 0D tunnelling. This is made clearer by plotting the resonant voltage as a function of the

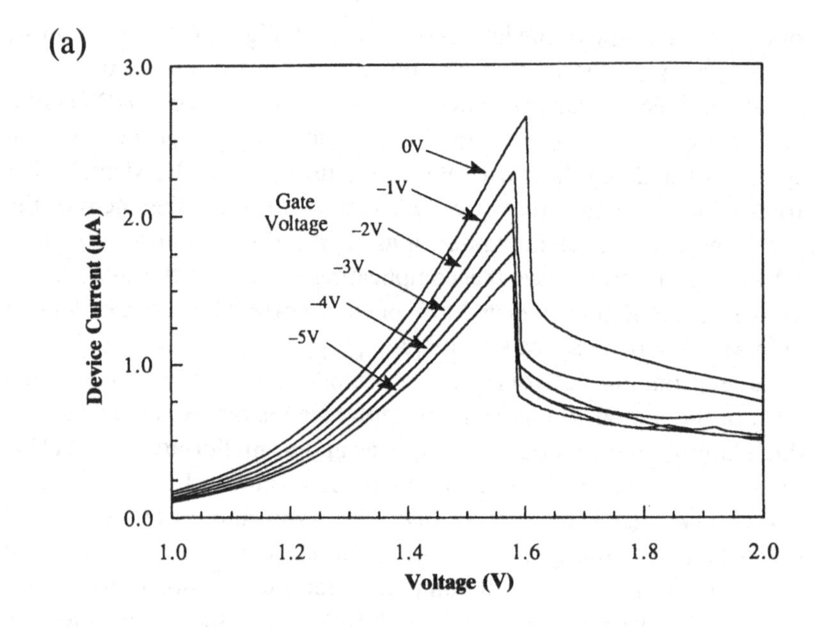

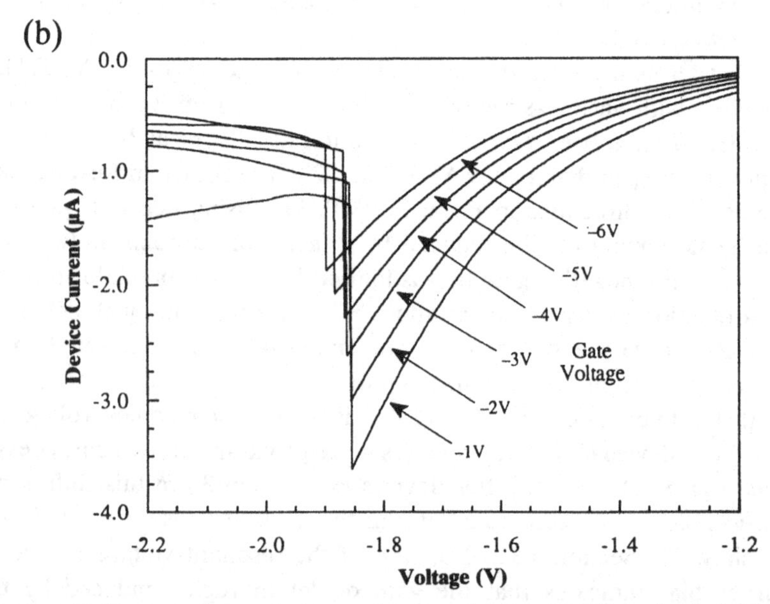

Figure 6.22 Characteristics of a 3 μm diameter device at 4.2 K on Material 2 in (a) forward bias and (b) reverse bias. The gate bias is referred to the potential of the bottom contact. Gate leakage currents of order 100 pA were measured for this device. After Goodings *et al.* [35], with permission.

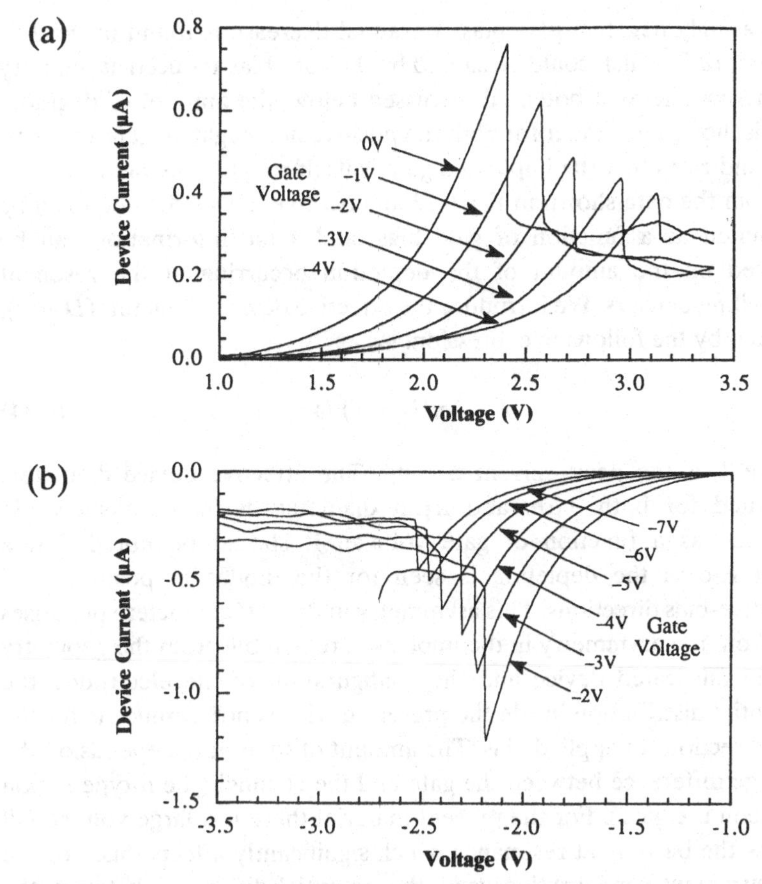

Figure 6.23 Characteristics for a 2 μm device at 4.2 K on Material 2 (a) in forward bias and (b) reverse bias. After Goodings *et al.* [35], with permission.

applied voltage relative to the turn-on voltage of the p–n-junction, as discussed in Section 6.3.2.

The symmetry of the lateral confinement potential around the barriers which is expected for the VARTD becomes evident through the observed *P*/*V* current ratios. In the case of the surface-gated VARTDs referred above, Beton *et al.* [41] found that the *P*/*V* ratio in one bias remained roughly constant while in the other bias the ratio diminished rapidly with applied gate voltage. This has been attributed to a gate voltage dependent asymmetry in the lateral confinement potential, resulting from the gate electrode being located on the surface. In the present case, however, the *P*/*V* ratio shows similar characteristics in both the forward and reverse directions, as seen in Fig. 6.22. With an increasingly negative gate bias, a gradual decrease is found in the *P/V* current ratio which could be caused by the gate-bias-induced asymmetry of the confinement potential discussed below; the amount of degradation is, however, similar for both the positive and negative gate voltages. This indicates that the implanted gate is itself roughly symmetrical.

From the data shown in Fig. 6.22 an effective device diameter can be estimated as a function of gate bias, and some information can be derived on the amount of the depletion occurring at the resonant tunnelling barriers. We introduce the effective device diameter, $(D - x)$, defined by the following expression:

$$I = I_0 \pi (D - x)^2/4 \tag{6.34}$$

where I_0 is the peak current density. The effective device diameters obtained for both 3 μm and 2 μm diameter devices are shown in Fig. 6.24 as a function of gate voltage. It should be noted that a difference in the depletion is seen for the diodes in positive- and negative-bias directions. This asymmetry in the *I–V* characteristics arises not from the asymmetry in the implanted region but from the geometry of the fabricated device and the configuration of the electrodes: the potential distribution inside the present device is not symmetric for the two directions of applied bias. The amount of squeezing depends on the voltage difference between the gate and the channel (the n-type region between the gate). For the present material there is a large voltage fall across the barriers at resonance which significantly affects this. For the measurement configuration used, the potential difference between the gate and the lower part of the channel remains constant in both forward and reverse bias, whereas the potential difference between the gate and the upper part of the channel alters by approximately twice the resonance voltage, with the greatest squeezing occurring in forward bias. For the 3 μm device given here, this is about 3.4 V. Assuming that the implant is symmetrical we would therefore expect that the effective device diameter for the 3 μm diode in forward and reverse bias would be identical but displaced by a gate voltage of about 3.4 V. The results given in Fig. 6.24 are indeed consistent with this.

6.3.2 Three-dimensional confinement effects on tunnelling characteristics

Now let us consider smaller devices in which we would expect to see the

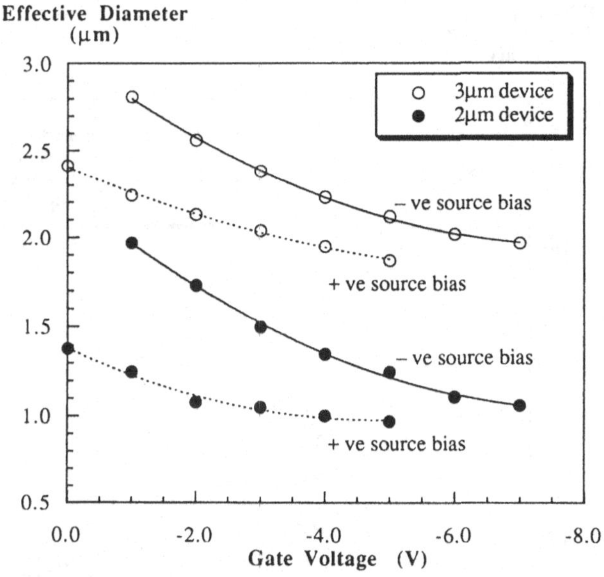

Figure 6.24 Effective electrical device diameter for 2 μm and 3 μm diameter top pads in forward and reverse bias at 4.2 K. After Goodings *et al.* [35], with permission.

effects of 3D quantisation on the characteristics. VARTDs with top pads of 1.8, 1.4, 1.2 and 1.0 μm in diameter have been fabricated using Material 3. Among these devices, the smallest one (1.0 μm diameter) has been found to be completely pinched off even at zero gate bias. This is in contrast to the results obtained using surface-gated VARTDs which had open channels even for sizes as small as 0.4 μm. This fact may imply that the implanted gate structure achieves a wider depletion region, which enables us to observe an important regime using relatively large devices. Therefore VARTDs with sizes between 1.0 and 2.0 μm are expected to be appropriate for the present purpose. Figure 6.25 shows the effective electrical diameters calculated from the measured peak currents for a range of devices and the material current density. Of course, for such small devices the concept of effective area becomes less well defined as the non-sharp nature of the confinement becomes more important. The results for three circular devices and one strip device (discussed in Section 6.4) are shown, with the circular devices having top pad diameters of 1.8, 1.4 and 1.2 μm. The devices show a similar dependence of the effective diameter on gate bias to one another, although some variations are seen which can be attributed to minor

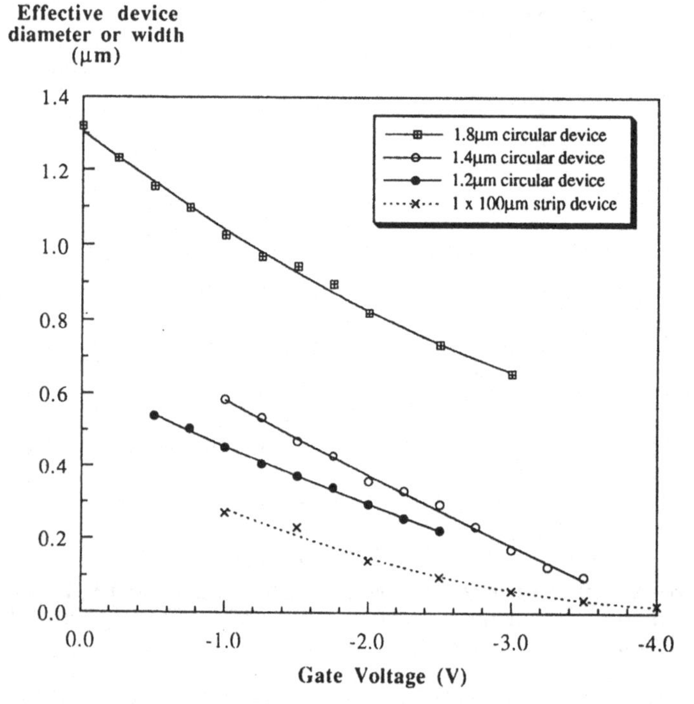

Figure 6.25 Summary of effective device sizes for small-area devices fabricated on Material 3. Three results are shown for circular devices and one for a strip device, discussed in Section 6.4. All curves correspond to measurements at 4.2 K. After Goodings *et al.* [35], with permission.

differences in the fabrication. A trend is also seen in the 'zero gate bias' depletion size, which ranges from about 200 nm for the 1.8 μm device to about 300 nm for the 1.2 μm device. This can again be attributed to the differing device geometries.

Figure 6.26 shows the characteristics of a 1.4 μm device at various gate biases: this device is pinched off at a gate bias of about −3.5 V. The *I–V* curves are plotted with a constant displacement along the current axis for clarity. Only the reverse-bias characteristics are shown since in the positive bias the channel is pinched-off even at zero gate bias, probably due to the asymmetry of squeezing discussed in the preceding section. This device, in contrast to the larger devices, is found to have *I–V* characteristics that are rich in structure: six small current peaks can be seen in total and their positions strongly depend on the gate bias. Also, the increase in the resonance voltage with increasing gate bias is much larger than that for the 3 μm device. The *I–V* curves are given in an

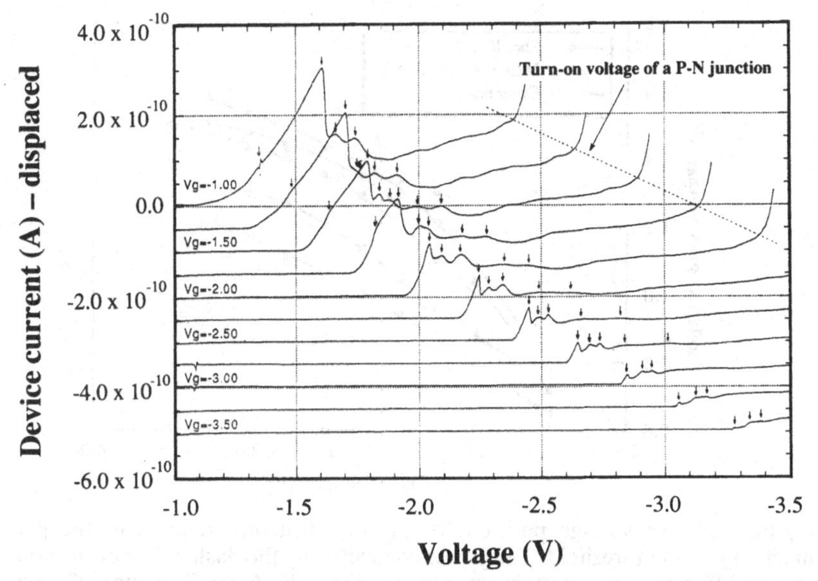

Figure 6.26 Characteristics of a 1.4 μm diameter device on Material 3 at 4.2 K in reverse bias. The main resonance peak is seen to sharpen and extra structure occurs in the valley current as the gate bias is driven increasingly negative. Successive curves have been displaced for clarity. After Goodings [33], with permission.

extended voltage region, and so a steep increase in the current is seen at a voltage above the resonance voltage which results from the turn-on of the gate-to-emitter p–n-junction. The shift of the current threshold for the turn-on of the p–n-junction is found to be almost exactly the same as the increase in the applied gate bias, as expected. The resonant peak voltage increases more slowly than the turn-on voltage for smaller gate biases, but starts to follow it as the gate bias is increased.

The situation becomes much clearer by calibrating the observed fine structure relative to the p–n-junction turn-on voltages. Due to the unknown amount of depletion region on the collector side of the barriers, the absolute voltages of the observed peaks are not meaningful. However, the p–n-threshold relates the emitter voltage to the gate voltage and so can be used as a reference point.

Figure 6.27 shows the gate voltage dependence of all six peak voltages. Except for the first current shoulder, indicated by a broken line, the graph appears to show the splitting of levels up to a gate voltage of about −2.0 V followed by a common shift in the relative peak positions. The shoulder indicated by the broken line is reminiscent of

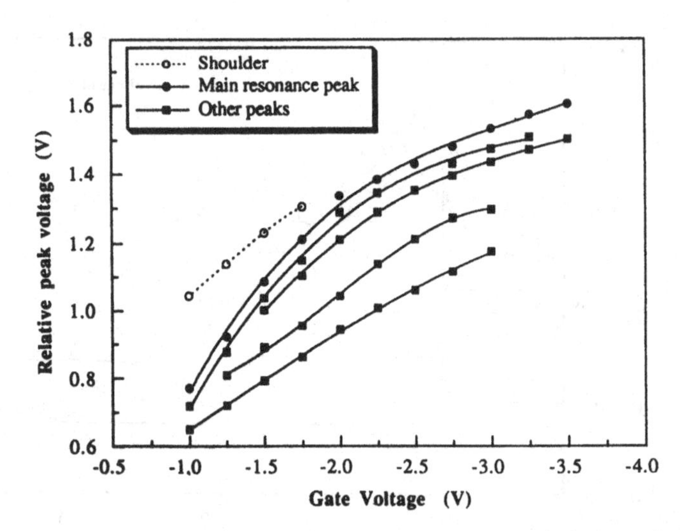

Figure 6.27 The voltage positions of the fine structure relative to the p–n-junction turn-on threshold. The points connected by the dashed line correspond to a shoulder seen on the main resonance in Fig. 6.26. After Goodings [33], with permission.

the characteristics produced by the 3D emitter states discussed in Section 4.1. As mentioned there, the low-doped emitter region gives rise to a triangular potential well next to the emitter barrier and an electrostatic bump forms slightly further away. The main tunnelling occurs from the localised 2D states in the emitter well, but the small contribution from non-localised 3D states results in broadening of the resonance and in even a separated resonance peak of its own under the right conditions. If this is the case here, the increasingly negative gate bias will act to raise the electrostatic bump still further, thereby reducing the contribution to the non-localised states and sharpening the resonance.

Lateral confinement potential profiles calculated for the 1.4 μm device using a classical simulation [27] are shown in Fig. 6.28. As the gate bias is increased, the lateral confinement well becomes progressively narrower until at about −2.0 V a minimum size is reached. Beyond this bias voltage, the effect is simply to shift the whole well upwards in energy without altering its shape. Such a result appears consistent with observations of splitting of levels followed by a uniform shift: once the minimum-well size is reached, the energy-level splitting between quantised levels no longer increases. The energy-level splitting evaluated at $V_g = -1.5$ V is about 5.5 meV, which is sufficiently large to

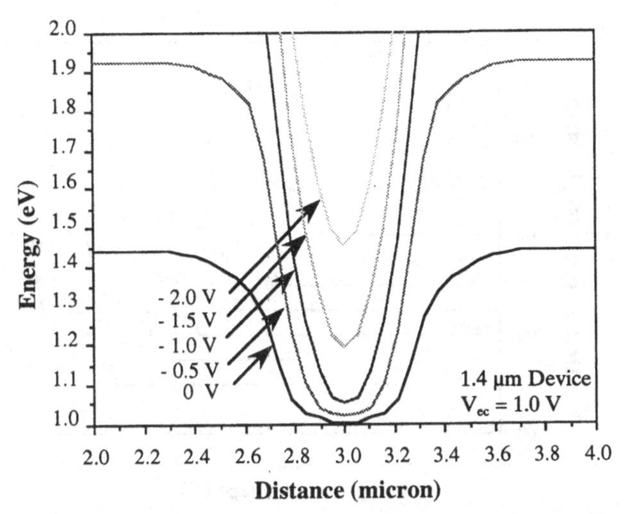

Figure 6.28 Lateral confinement potential profile calculated for the 1.4 μm device at various negative gate biases.

be resolved at 4.2 K. Using the energy level associated with large-area RTD resonance to calibrate the measured voltages to the energies in the well, this predicts a measured splitting of about 150 mV, which is again consistent with Fig. 6.27.

Similar results have been observed for the 1.2 μm device, and these are shown in Fig. 6.29(a) and (b). In Fig. 6.29(b), all of the curves become flatter at the higher gate voltages since the system exceeds the squeezing limit at a lower gate voltage than does that for the 1.4 μm device. These results are consistent with the above discussion on the fine structure and may indicate that the fine structure obtained here is attributable to 3D quantisation.

6.3.3 Tunnelling through single-impurity states

The VARTDs exhibit another interesting feature at their current threshold. Figure 6.30 shows the *I–V* characteristics of the 2 μm device around threshold in forward bias (note the difference in the scale for the longitudinal axis from that in Fig. 6.23). This shows a series of plateau-like structures at threshold. Similar structures are seen for many devices with different sizes: *I–V* curves of the 4 μm device are shown in Fig. 6.30 for both bias directions: (b) in forward and (c) reverse bias. It should be noted that the structure seen differs in the forward- and reverse-bias

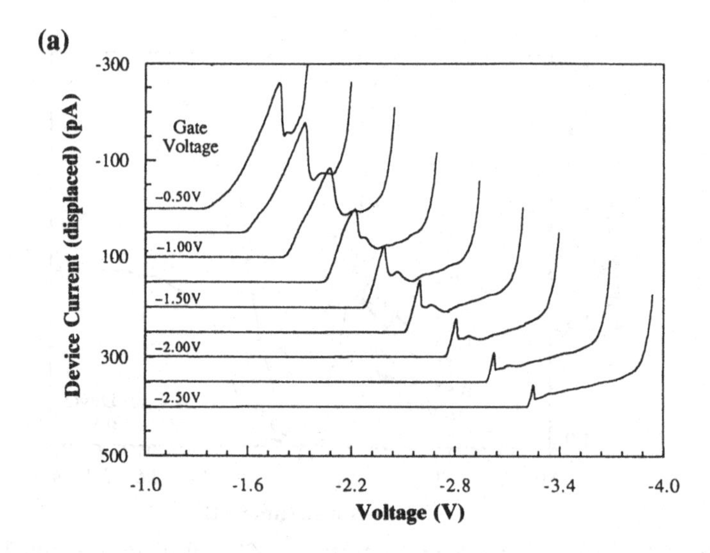

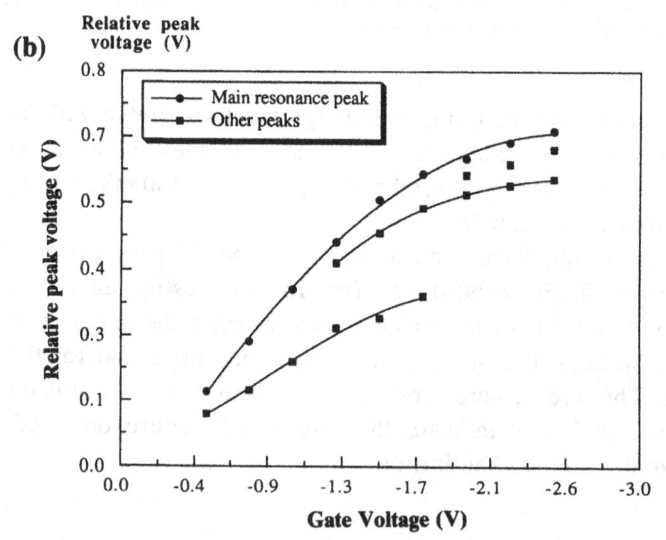

Figure 6.29 (a) Characteristics of a 1.2 μm diameter device on Material 3 at 4.2 K in reverse bias. The main resonance peak is seen to sharpen and extra structure occurs in the valley current. Part (b) plots the peak voltage positions relative to the p–n-junction turn-on threshold. (After Goodings [33], with permission.)

directions. The gate-bias dependencies of the characteristics are dependent on device size: for the 2 μm device the position of the plateaux is dependent on gate voltage, while for the larger devices this is not the case. This gate-bias dependence seen for the 2 μm device, however, is

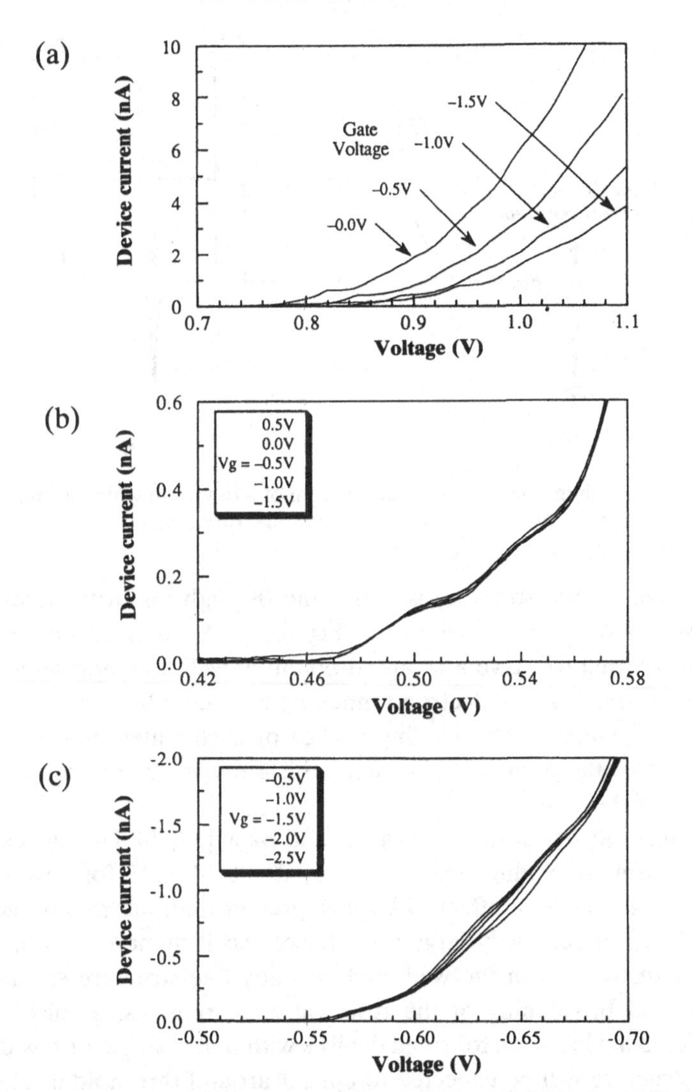

Figure 6.30 Fine structure observed near threshold for VARTDs at 4.2 K: (a) for a 2 μm diameter device in forward bias, (b) for a 4 μm diameter device in forward bias, and (c) for the same 4 μm diameter device in reverse bias.

thought to be caused by the same mechanism as the peak voltage shift explained in the preceding section. Thus, in general, the present fine structure is expected to have a very small gate-bias dependence.

Very similar results have been reported by Dellow *et al.* [36]–[38] for their surface-gated VARTDs. They pointed out that one possibility for

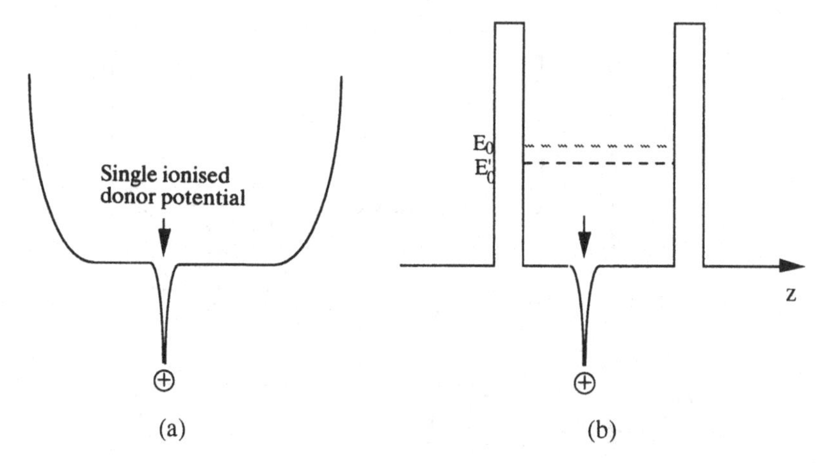

Figure 6.31 Schematic energy band-diagram of a device in which a single ionised donor is placed: (a) in lateral and (b) in vertical directions.

the origin of the structure is tunnelling through impurity states in the active region of the device (see Fig. 6.31). An ionised donor in the quantum well will give a localised potential well and associated bound states through which electron tunnelling can occur for biases below the threshold voltage. The binding energy of such states depends on the position of the donors in the well, with a maximum for donors in the centre of the well.

Using modelled results by Greene and Bajaj [39], [40] we can estimate the maximum binding energy to be about 14 meV for this system, indicating that these effects are much greater than thermal broadening at 4.2 K. For relatively large-area devices such numerous donors randomly distributed in the well wash out any fine structure so that only collisional broadening of the resonant state remains, as discussed in Section 3.3. However, for small devices with only a single or few donors, fine structure will be expected to appear around threshold as electrons tunnel through the local quasi-bound state associated with the ionised donors.

The total number of donor states expected in the quantum well can be estimated from the background doping density and the active device volume (see Table 6.1). Taking a background doping of 10^{14} cm^{-3} this gives an estimate of 1–5 donor sites in the three device sizes considered here, which is consistent with the single- or few-electron tunnelling picture. The observed asymmetry of the forward and reverse characteristics is also consistent with this picture, as a single impurity is placed

Table 6.1. Details of the first plateau currents, dimensions of the active regions and associated dwell times derived using eqn (6.35) assuming that only single electron tunnelling is involved

Top pad diameter (μm)	First plateau current (pA)	Calculated dwell time (ps)	Active device area (m^2)	Active device volume (m^3)
2.0	500	320	1.0×10^{-12}	0.6×10^{-20}
3.0	250	640	3.6×10^{-12}	2.1×10^{-20}
4.0	150	1500	7.8×10^{-12}	4.6×10^{-20}

completely randomly in the well. Unless the impurity is located at the centre of the active area, some asymmetry is always expected for the characteristics.

Let us consider an extreme case in which only one electron at a time can contribute to the conduction. The tunnelling current associated with a single electron tunnelling through a single ionised donor state can be roughly estimated by the following expression:

$$I = e/\tau_d \tag{6.35}$$

which corresponds to eqn (2.41) for large-area devices. By assuming this model and using the observed currents of the first plateaux, values of the dwell time, τ_d, are obtained for three VARTDs which are listed in Table 6.1. The value of τ_d obtained for large-area devices using eqn (2.41) and the charge density in the well, σ_w, estimated from the magneto-conductance measurements (see Section 4.3.2) is about 500 ps. The above values of τ_d calculated by assuming the single-electron tunnelling model (eqn (6.35)) are fairly close to the value derived for large-area devices. This fact may indicate that the observed plateaux are attributable to single-electron tunnelling through few ionised donor states. The electron dwell time varies with device size, as seen in Table 6.1, but this could be accounted for by the difference in the effective electric field across the double-barrier structure.

Detailed analysis of single-impurity-related tunnelling requires the 3D *S*-matrix simulation introduced in Section 6.2. The attractive potential due to an ionised single donor, $V_{IM}(x,y,z)$, is simply modelled by using a 3D delta function:

$$V_{IM}(\mathbf{r}) = -V_0\delta(\mathbf{r} - \mathbf{r}_0) \tag{6.36}$$

which is introduced into eqn (6.2). Scattering matrix elements (eqn (6.15)) are calculated by using the lateral wavefunctions which are obtained from the 2D Schrödinger equation (eqn (6.4)). A uniform lateral confinement has been adopted for the present analysis, taking into account that a virtually flat confinement is achieved in the VARTD. The current analysis has been performed for a relatively large-area device on Material 2 (a square potential well of 200 nm in diameter is assumed) in which contributions from different lateral modes are hardly separable. Thus the *I–V* characteristics show only a single current peak, even at zero temperature, if no ionised impurities are introduced.

Current–voltage characteristics for the RTD calculated near the threshold are shown in Fig. 6.32(a) and (b) where a single ionised donor is assumed near the centre of the dot.

The total transmission probability through an RTD in which a single ionised donor is located at the centre of the structure is shown in Fig. 6.33(a)–(c), where Fig. 6.33(a) shows characteristics calculated at threshold ($V = 165.0$ mV). A single transmission peak is found at a lower energy than, and isolated from, a series of other peaks which are well overlapped leading to a broad transmission band. This transmission peak is found to give rise to a small current peak, as shown in Fig. 6.33(b). The 3D probability density calculated for the first-mode incident wave at an energy of 184.155 meV under an applied bias of 177.5 mV is shown in Fig. 6.34: it can clearly be seen that the electron wave is now localised around the single ionised donor despite the widely spread nature of the incident wave at the emitter edge. After this single peak falls below the conduction-band edge in the emitter, a group of other transmission peaks falls to the Fermi sea leading to the main current peak (Fig. 6.33(c)). These preliminary results demonstrate that a single ionised donor placed in a quantum well indeed gives rise to a small but observable current peak near the threshold of the main current peak.

6.4 One-dimensional (1D) resonant tunnelling diodes

Before going to the final section, let us see the results obtained from 1D VARTDs in which the top pad, rather than being circular, takes the form of a long, thin rectangle. The motivation behind this is to increase the total device current whilst still maintaining a large amount of control over the device area.

The characteristics of a 1 μm × 100 μm device are shown in Fig. 6.35.

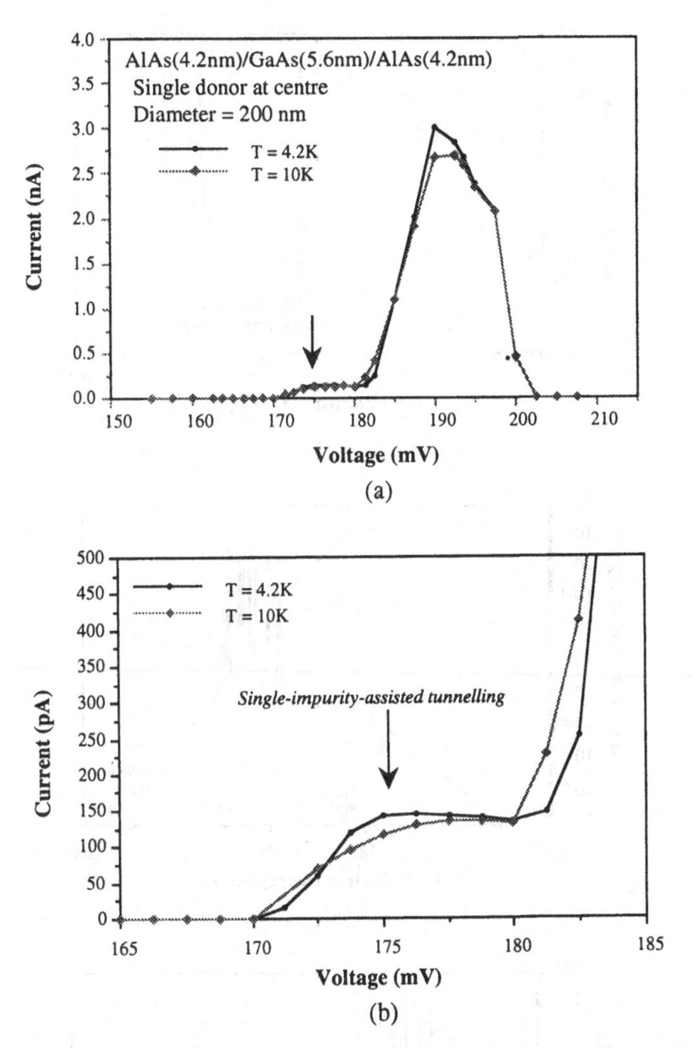

Figure 6.32 *I–V* characteristics of a large-area RTD with a single ionised donor placed at the centre of the resonant tunnelling structure calculated at 4.2 K (solid line) and 10 K (dotted line): (a) full characteristics and (b) detail around the threshold.

As expected for such a device in which one dimension is small, a large amount of control is achieved over the current. The effective device area versus gate voltage has already been shown in Fig. 6.25. However, we see that, in contrast to the smooth reduction in peak current of the small circular devices, for these diodes the resonance peak appears to break up into numerous smaller peaks at different voltages. Figure 6.35(b)

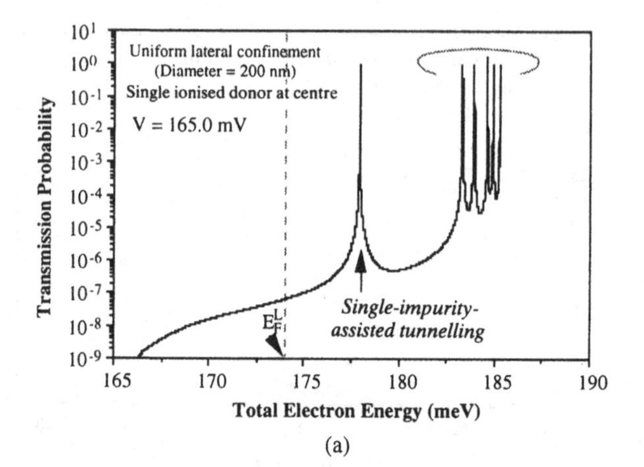
Uniform lateral confinement
(Diameter = 200 nm)
Single ionised donor at centre
V = 165.0 mV
Transmission Probability
Single-impurity-assisted tunnelling
E_F^L
Total Electron Energy (meV)
(a)

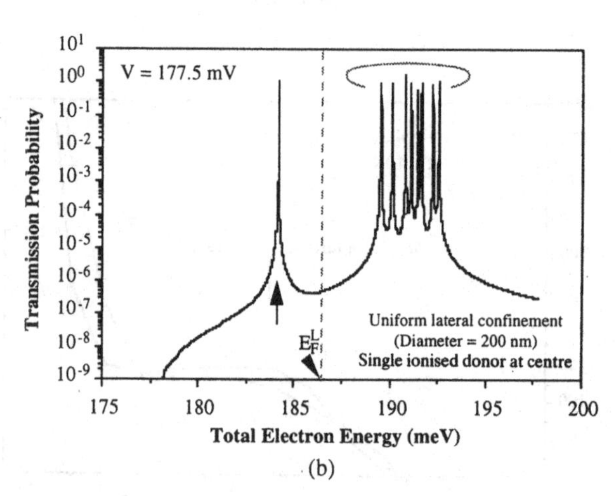
V = 177.5 mV
Transmission Probability
Uniform lateral confinement
(Diameter = 200 nm)
Single ionised donor at centre
E_F^L
Total Electron Energy (meV)
(b)

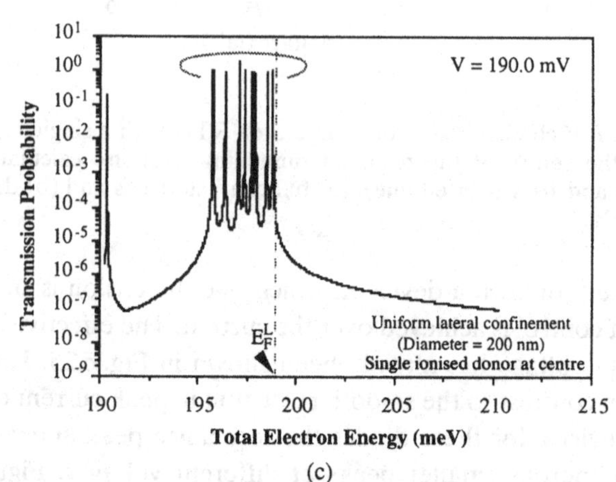
V = 190.0 mV
Transmission Probability
Uniform lateral confinement
(Diameter = 200 nm)
Single ionised donor at centre
E_F^L
Total Electron Energy (meV)
(c)

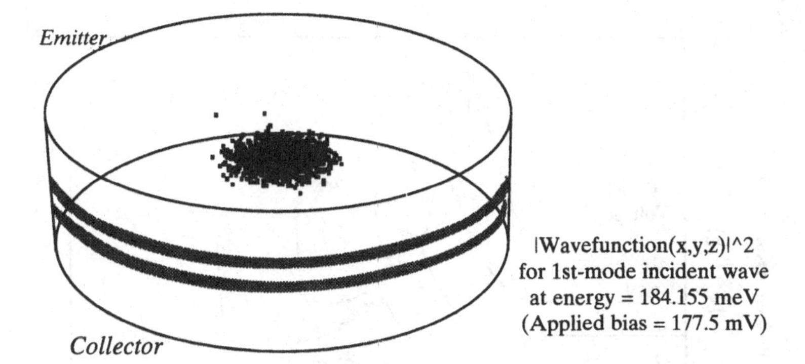

Figure 6.34 Visualised probability density of electrons calculated for the first mode incident wave with an energy of 184.155 meV (at a single impurity related resonance) under an applied voltage of 177.5 mV (see Fig. 6.33(b)). An ionised donor is placed at the centre of the structure.

shows a more detailed view for a smaller range of gate biases. On this scale the structure starts to appear less random, with trends being seen for the current peaks on successive curves. The heights of the current steps have been measured and are plotted versus the device bias in the inset to Fig. 6.35(b), from which it is apparent that the current steps fall into multiples of some unit current step which becomes smaller with increasing voltage. This holds for a range of gate biases. A value for the unit-step can be derived from the inset. Taking as our example the bias $V = -1.8$ V, the graph gives a unit current step of order 500 pA. This current, ΔI, can then be related to the number of electrons associated with the step, n, and their dwell time in the quantum well, τ_d, by:

$$\Delta I = \frac{ne}{\tau_d} \tag{6.37}$$

Using a τ_d of 100 ns derived for Material 3 from measurements on large-area devices, this gives the number of electrons associated with the unit current step as about 300, which clearly eliminates the possibility of the steps being due to single-electron effects.

Figure 6.33 Transmission probability through a large-area RTD with a single ionised donor placed at the centre of the resonant tunnelling structure calculated at three different biase: (a) at the current threshold ($V = 165.0$ mV), (b) below the current peak ($V = 177.5$ mV) and (c) at the main current peak ($V = 190.0$ mV). An arrow indicates the transmission peak caused by the single ionised donor potential.

(a)

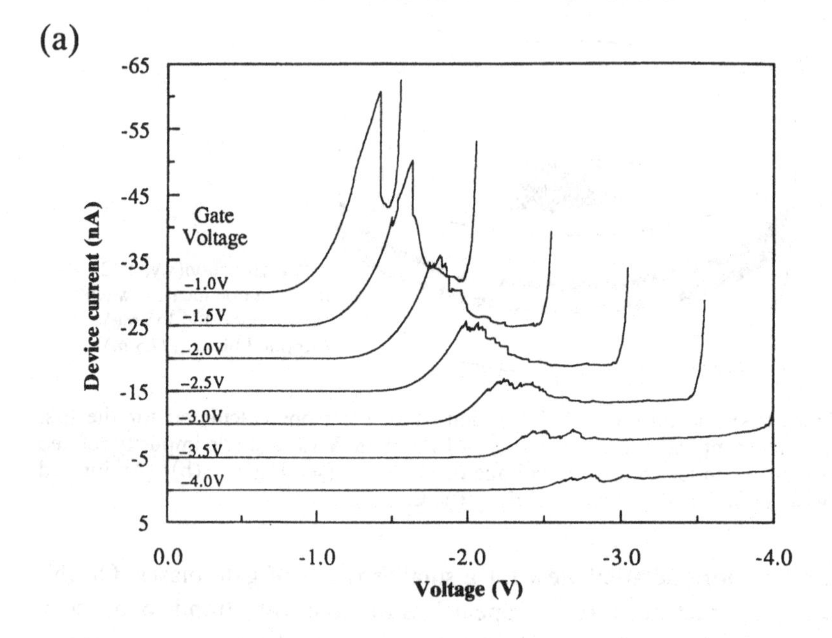

(b)

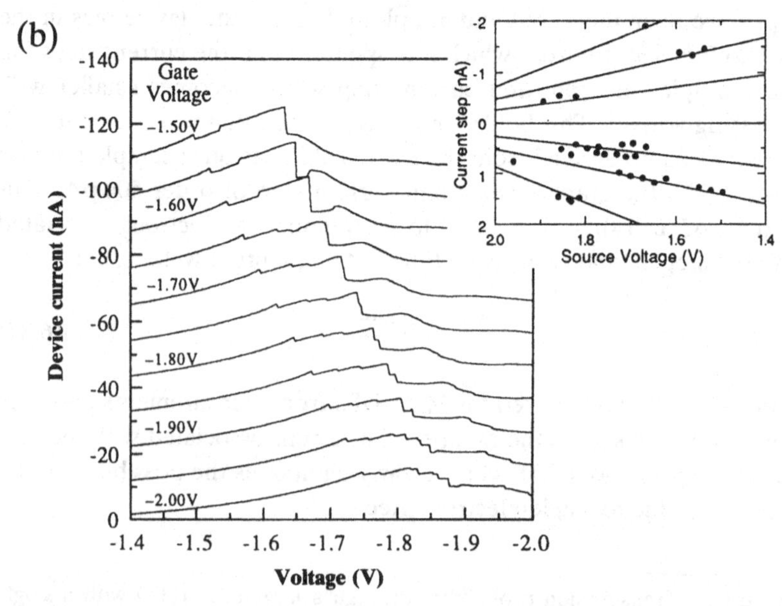

Figure 6.35 (a) Characteristics of a $1 \times 100\,\mu m^2$ strip device at 4.2 K. The roughness seen is real and repeatable, part (b) shows a section in more detail in which the trends between the curves begin to become apparent. The inset shows a plot of the height of the current steps versus the device voltage in (b). The heights of the steps appear to fall roughly in quantised values.

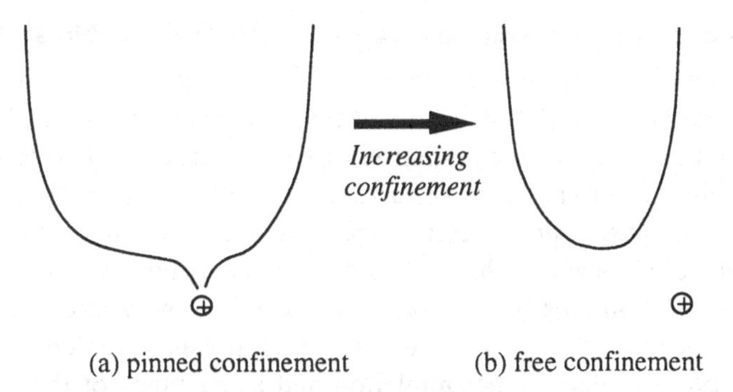

Figure 6.36 Sketch of lateral confinement potential pinned at a donor state in the active region (a). Eventually, the confinement potential will overcome the pinning (b), producing a step in current.

Instead, the proposed mechanism for these characteristics involves the pinning of the confinement potential by donor sites in the active region of the device. This is illustrated in Fig. 6.36. The lateral confinement is a non-trivial function of the source and gate potentials, but in general increases as either the gate voltage or the difference between the source and gate voltages increases. As this occurs, the lateral confining potential indicated in Fig. 6.36 will become narrower. If, however, there is a donor site in the active region, as shown, one wall of the confining potential may become pinned. Eventually, the applied potential will become large enough to overcome the pinning and a sudden step in confining potential will occur, with a corresponding step in the observed current. The magnitude of this current step will depend upon the step change in the confining potential and so can correspond to many electron tunnelling events, as observed. If two or more pinning centres are close enough to be coupled, then multiples of the unit current step or bistable regions may be seen, also as observed.

The number of steps seen in the *I–V* characteristics increases as the gate voltage becomes increasingly negative. Actually, the important varying quantity here is the difference between the gate and source voltages and we see that this changes over a greater range for the higher gate biases. An estimate of the number of donor sites in the active volume yields, for a background doping of 10^{14} cm^{-3}, 30 donors, which is the correct order for the number of current steps observed in the characteristics for higher gate voltages.

6.5 Interplay of resonant tunnelling and Coulomb blockade

In this final section we briefly study the effects of space charge on the tunnelling properties of 0D resonant tunnelling structures. In Chapter 4 we saw that electron accumulation in the quantum well of large-area RTDs gives rise to an intrinsic current bistability through different self-consistent fields. Typical electron accumulation is found of the order of 10^{11} cm^{-2} at resonance, which gives, for instance, about 10^5 electrons for an RTD with an area of 100 μm^2. This phenomenon can be found for a wide range of device areas as the self-consistent fields are determined by the amount of electron accumulation and capacitance of the resonant tunnelling structure. For 0D RTDs, however, a more striking effect is expected for the charging of the quantum dot: the electrostatic potential caused by a *single electron* in the 0D structure may affect the resonant tunnelling process.

The charging energy of the 0D structure through a single electron is roughly evaluated in the following way, although its proper estimation obviously requires 3D self-consistent simulation. Let us consider a simple cylindrical quantum pillar with a sufficiently small radius, r_s, to achieve a nearly parabolic lateral confinement, $V_{LC}(r)$:

$$V_{LC}(r) = \frac{1}{2}m^* \omega_{LC}^2 r^2 \tag{6.38}$$

which produces an equal separation between lateral quantised levels of $\epsilon_{LC} = \hbar\omega_{LC}$. The total capacitance of the 0D double-barrier structure, C_{tot}, may be calculated using a formula for two sequential parallel-plate capacitors; emitter and collector barrier capacitors, C_{eb} and C_{cb}, with an area of πr_c^2 (r_c is an effective channel radius)

$$\frac{1}{C_{tot}} = \frac{1}{C_{eb}} + \frac{1}{C_{cb}} \tag{6.39}$$

The charging energy is then simply given by

$$\epsilon_c = \frac{e^2}{2C_{tot}} \tag{6.40}$$

The calculated charging energy is shown in Fig. 6.37 as a function of r_s in comparison with the corresponding lateral quantisation energy. An $Al_{0.3}Ga_{0.7}As/GaAs$ symmetric double-barrier structure with a barrier thickness of 10 nm is used for the present calculation. The surface potential, V_s, and depletion width, $r_s - r_c$, are assumed to be 0.7 eV and 140 nm respectively. It is found that the single-electron charging energy

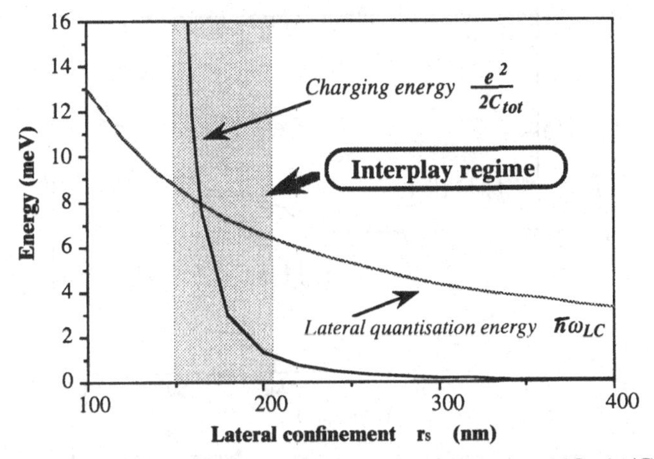

Figure 6.37 Single-electron charging energy calculated for an AlGaAs/GaAs 0D RTD as a function of lateral confinement radius (solid line). The shaded region indicates an interplay regime with lateral quantisation effect, shown by the grey line.

becomes a few milli-electron volts for r_s less than 200 nm, that is, r_c less than 60 nm. At low temperatures this charging energy is much larger than thermal energy, k_BT, and thus results in a characteristic *Coulomb blockade* (CB) phenomenon in its I–V curve [42]. There is the possibility of being able to observe an interplay of the single-electron charging effect with the lateral quantisation effect in this regime, and this has attracted much interest recently.

Guéret *et al.* [43] first observed current plateaux near the current threshold using Schottky-gated squeezable 0D RTDs with a diameter of 400 nm and a symmetric $Al_{0.13}Ga_{0.87}As$(13.5 nm)/GaAs(7 nm)/$Al_{0.13}Ga_{0.87}As$(13.5 nm) double-barrier structure. The I–V characteristics of this device at 40 mK exhibit a few current plateaux with a nearly constant magnitude of 1 nA over a wide range of gate voltages. A slight current modulation was found to be superposed on the current step which Guéret *et al.* attributed to an interplay of the Coulomb blockade with the lateral quantisation.

More experimental evidence has been presented in a clearer manner by Tewordt *et al.* [44] and Su *et al.* [45], [46] using similar 0D RTDs with asymmetric double-barrier structures. As we saw in Section 4.3, the space-charge build-up is enhanced in asymmetric barrier structures when an opaque barrier is placed on the collector side. Tewordt *et al.* [44] used an etched two-terminal 0D RTD with an $Al_{0.33}Ga_{0.67}As$(10 nm)/GaAs(14 nm)/$Al_{0.33}Ga_{0.67}As$(7 nm) double-barrier

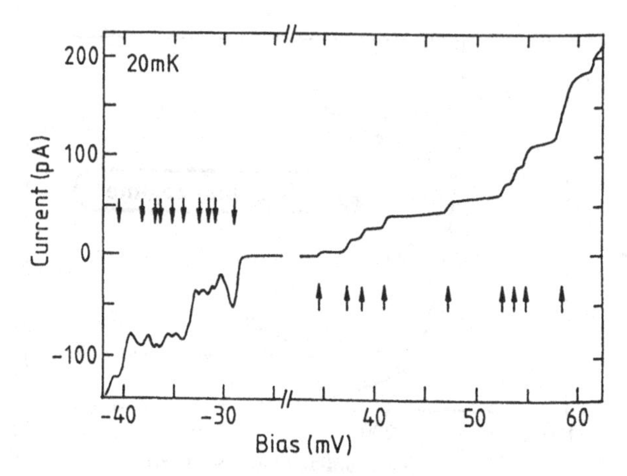

Figure 6.38 *I–V* characteristics of an asymmetric 0D RTD in the low-bias region. In forward bias a Coulomb staircase is seen, while the resonance current spikes are found in reverse bias which are attributable to 0D resonant tunnelling. After Tewordt *et al.* [44], with permission.

structure with thick low-doped (2×10^{16} cm^{-3}) n-GaAs layers on both sides. In this device the effect of lateral quantisation and that of the Coulomb blockade are expected to be separated in the two bias directions. The *I–V* characteristics for such 0D RTDs with a diameter of 1 μm measured at 20 mK in both bias directions are shown in Fig. 6.38. Tewordt *et al.* found sharp current steps attributable to Coulomb blockade when the transparent AlGaAs barrier is located on the emitter side (this is referred to as a forward-bias direction), while a more spikey *I–V* curve was observed in a reverse-bias direction which is due to 0D resonant tunnelling. The step-like increase in current (Coulomb staircase) in the positive direction is thought to correspond to the increase in the electron number accompanied by Coulomb blockade. Similar experimental results have also been reported by Su *et al.* [45], [46] using a different asymmetric $Al_{0.34}Ga_{0.66}As$(10 nm)/GaAs(9 nm)/ $Al_{0.36}Ga_{0.64}As$(11.5 nm) structure. The widths of the current steps correspond to the energy required to add one more electron to the quantum dot and thus should include information about the Coulomb interactions among few electrons. The first systematic studies of the step widths have been performed recently by Su *et al.* [45], [46] using magnetic fields perpendicular to the barriers, and evidence of the transition from the spin singlet ground state to a spin triplet state [47] has been demonstrated for the two-electron states.

6.6 References

[1] M. A. Reed, J. N. Randall, R. J. Aggarwal, R. J. Matyi, T. M. Moore and A. E. Westel, Observation of discrete electronic states in a zero-dimensional semiconductor nanostructure, *Phys. Rev. Lett.*, **60**, 535, 1988.

[2] M. A. Reed, J. N. Randall and J. H. Luscombe, Non-equilibrium quantum dots: transport, *Nanotechnology*, **1**, 63, 1990.

[3] S. Tarucha, Y. Hirayama, T. Saku and T. Kimura, Resonant tunneling through one- and zero-dimensional states constricted by $Al_xGa_{1-x}As$/GaAs/$Al_xGa_{1-x}As$ heterojunctions and high-resistance regions induced by focused Ga ion-beam implantation, *Phys. Rev.*, **B41**, 5459, 1990.

[4] Bo Su, V. J. Goldman, M. Santos and M. Shayegan, Resonant tunneling in submicron double-barrier heterostructures, *Appl. Phys. Lett.*, **58**, 747, 1991.

[5] S. Tarucha and Y. Hirayama, Magnetotunneling in a coupled two-dimensional–one-dimensional electron system, *Phys. Rev.*, **B43**, 9373, 1991.

[6] M. W. Dellow, P. H. Benton, M. Henini, P. C. Main, L. Eaves, S. P. Beaumont and C. D. W. Wilkinson, Gated resonant tunnelling devices, *Electron. Lett.*, **27**, 134, 1991.

[7] P. Guéret, N. Blanc, R. German and H. Rothuizen, Confinement and single-electron tunneling in Schottky-gated, laterally squeezed double-barrier quantum-well heterostructures, *Phys. Rev. Lett.*, **68**, 1896, 1992.

[8] S. Tarucha, Y. Tokura and Y. Hirayama, Resonant tunneling of three-dimensional electrons into degenerate zero-dimensional levels, *Phys. Rev.*, **B44**, 13815, 1991.

[9] A. Groshev, Single electron trapping in ultrasmall double barrier semiconductor structures, *Twentieth International Conference on the Physics of Semiconductors, Thessaloniki*, World Scientific, p. 1238, 1990.

[10] A. Groshev, T. Ivanov and V. Valtchinov, Charging effects of a single quantum level in a box, *Phys. Rev. Lett.*, **66**, 1082, 1991.

[11] M. P. Stopa, Charging energy and collective response of a quantum dot resonant tunneling device, *Surface Science*, **263**, 433, 1991.

[12] G. W. Bryant, Resonant tunneling in zero-dimensional nanostructures, *Phys. Rev.*, **B39**, 3145, 1989.

[13] G. W. Bryant, Understanding quantum-box resonant-tunneling spectroscopy: Fine structure at Fermi-level crossings, *Phys. Rev.*, **B44**, 3782, 1991.

[14] G. W. Bryant, Nonadiabatic transport through quantum dots, *Phys. Rev.*, **B44**, 12837, 1991.

[15] M. Luban, J. H. Luscombe, M. A. Reed and D. L. Pursey, Anharmonic oscillator model of a quantum dot nanostructure, *Appl. Phys. Lett.*, **54**, 1997, 1989.

[16] J. R. Barker, Theory of quantum transport in lateral nanostructures, *Nanostructure physics and fabrication*, *Proceedings of the International Symposium, Texas*, Academic Press, p. 253, 1989.

[17] F. Sols, M. Macucci, U. Ravaioli and K. Hess, Theory of a quantum modulated transistor, *J. Appl. Phys.*, **66**, 3892, 1989.

[18] J. A. Brum, Superlattice effects in quantum dots, *Twentieth International Conference on the Physics of Semiconductors, Thessaloniki*, World Scientific, p. 511, 1990.

[19] S. E. Ulloa, E. Castano and G. Kirczenow, Ballistic transport in a novel

one-dimensional superlattice, *Phys. Rev.*, **B41**, 12350, 1990.

[20] Z. L. Ji and K. F. Berggren, Numerical study of ballistic conductance in parallel configuration, *Semicond. Sci. Technol.*, **6**, 63, 1991.

[21] A. Weisshaar, J. Lary, S. M. Goodnick and V. K. Tripathi, Negative differential resistance in a resonant quantum wire structure, *IEEE Electron Device Lett.*, **EDL-12**, 2, 1991.

[22] A. Kumar and P. H. Bagwell, Resonant tunneling in a quasi-one-dimensional wire: Influence of evanescent modes, *Phys. Rev.*, **B43**, 9012, 1991.

[23] F. M. de Aguiar and D. A. Wharam, Transport through one-dimensional channels, *Phys. Rev.*, **B43**, 9984, 1991.

[24] A. Kumar, S. E. Laux and F. Stern, Electron states in a GaAs quantum dot in a magnetic field, *Phys. Rev.*, **B42**, 5166, 1990.

[25] A. Kumar, Self-consistent calculations on confined electrons in three-dimensional geometries, *Surface Science*, **263**, 335, 1992.

[26] K. Nakasato and R. J. Blaikie, The effect of mode coupling on ballistic electron transport in quantum wires, *J. Phys: Condens. Matter*, **3**, 5729, 1991.

[27] H. Mizuta, K. Yamaguchi, M. Yamane, T. Tanoue and S. Takahashi, Two-dimensional numerical simulation of Fermi-level pinning phenomena due to DX centers in AlGaAs/GaAs HEMTs, *IEEE Trans. Electron Device*, **ED-36**, 2307, 1989.

[28] W. E. Spicer, P. W. Chye, P. R. Skeath, C. Y. Su and I. Lindau, New and unified model for Schottky barrier and III–V insulator interface states formation, *J. Vac. Sci. Technol.*, **16**, 1422, 1979.

[29] See, for example, W. H. Press, B. P. Flannery, S. A. Teukolsky and W. T. Vetterling, 1986, *Numerical Recipes, The Art of Scientific Computing* (Cambridge University Press), Chapter 11.

[30] H. Mizuta, C. J. Goodings, M. Wagner and S. Ho, Three-dimensional numerical analysis of multi-mode quantum transport in zero-dimensional resonant tunnelling diodes, *J. Phys.: Condens. Matter*, **4**, 8783, 1992.

[31] U. Fano, Effects of configuration interaction on intensities and phase shifts, *Phys. Rev.*, **124**, 1866, 1961.

[32] C. J. Goodings, J. R. A. Cleaver and H. Ahmed, Variable-area resonant tunnelling diodes using implanted gates, *Electron. Lett.*, **28**, 1535, 1992.

[33] C. J. Goodings, Variable-area resonant tunneling diodes using implanted gates, PhD thesis, Cambridge University, 1993.

[34] C. J. Goodings, H. Mizuta, J. R. A. Cleaver and H. Ahmed, Electron confinement in variable-area resonant tunnelling diodes using in-plane implanted gates, *Surface Science*, **305**, 353, 1994.

[35] C. J. Goodings, H. Mizuta, J. R. A. Cleaver and H. Ahmed, Variable-area resonant tunnelling diodes using implanted in-plane gates, *J. Appl. Phys.*, **76**, 1276, 1994.

[36] M. W. Dellow, C. J. G. M. Langerak, P. H. Beton, T. J. Foster, P. C. Main, L. Eaves, M. Henini, S. P. Beaumont and C. D. W. Wilkinson, Zero dimensional resonant tunneling through single donor states, *Superlattices and Microstructures*, **11**, 149, 1992.

[37] M. W. Dellow, C. J. G. M. Langerak, P. H. Beton, T. J. Foster, P. C. Main, L. Eaves, M. Henini, S. P. Beaumont and C. D. W. Wilkinson, Single electron tunnelling through a donor state in a gated resonant tunnelling device, *Surface Science*, **263**, 438, 1992.

[38] M. W. Dellow, P. H. Beton, C. J. G. M. Langerak, T. J. Foster, P. C. Main,

L. Eaves, M. Henini, S. P. Beaumont and C. D. W. Wilkinson, Resonant tunneling through the bound states of a single donor atom in a quantum well, *Phys. Rev. Lett.*, **68**, 1754, 1992.
[39] R. L. Greene and K. K. Bajaj, Effect of magnetic fields on the energy levels of a hydrogenic impurity center in GaAs/$Ga_{1-x}Al_xAs$ quantum-well structures, *Phys. Rev.*, **B31**, 913, 1985.
[40] R. L. Greene and K. K. Bajaj, Far-infrared absorption profiles for shallow donors in GaAs-$Al_xGa_{1-x}As$ quantum-well structures, *Phys. Rev.*, **B34**, 951, 1986.
[41] P. H. Beton, M. W. Dellow, P. C. Main, T. J. Foster, L. Eaves, A. F. Jezierski, M. Henini, S. P. Beaumont and C. D. W. Wilkinson, Edge effects in a gated submicron resonant tunneling diode, *Appl. Phys. Lett.*, **60**, 2508, 1992.
[42] See, for example, B. L. Altshuler, P. A. Lett and R. A. Webb, 1991, *Mesoscopic Phenomena in Solids, Modern Problems in Condensed Matter Sciences*, Vol. 30 (North-Holland), Chapter 6.
[43] P. Guéret, N. Blanc, R. Germann and H. Rothuizen, Vertical transport in Schottky-gated, laterally confined double-barrier quantum well heterostructures, *Surface Science*, **263**, 212, 1992.
[44] M. Tewordt, L. Martin-Moreno, J. T. Nicholls, M. Pepper, M. J. Kelly, V. J. Law, D. A. Ritchie, J. E. F. Frost and G. A. C. Jones, Single-electron tunneling and Coulomb charging effects in asymmetric double-barrier resonant-tunneling diodes, *Phys. Rev.*, **B45**, 14407, 1992.
[45] Bo Su, V. J. Goldman and J. E. Cunningham, Single-electron tunneling in double-barrier nanostructures, *Superlattices and Microstructures*, **12**, 305, 1992.
[46] Bo Su, V. J. Goldman and J. E. Cunningham, Single-electron tunneling in nanometer-scale double-barrier heterostructure devices, *Phys. Rev.*, **B46**, 7644, 1992.
[47] M. Wagner, U. Merkt and A. V. Chaplik, Spin singlet-triplet oscillations in quantum dot helium, *Phys. Rev.*, **B45**, 1951, 1992.

Index

Printed in the United States
by Bookmasters

Printed in the United States
By Bookmasters